D1036274

The
original edition
of this book con-
tained oversized maps.
PDF files of these maps
can be downloaded from
www.elibron.com/maps

Francis Leopold McClintock

The Voyage of the 'Fox' in the Arctic Seas

A Narrative
of the
discovery of the fate
of
SIR JOHN FRANKLIN
and
his companions

Elibron Classics
www.elibron.com

Elibron Classics series.

ISBN 1-4021-6379-7 (paperback)
ISBN 1-4021-2164-4 (hardcover)

This Elibron Classics Replica Edition is an unabridged facsimile of the edition published in 1859 by John Murray, London.

THE 'FOX' STEAMING OUT OF THE ROLLING PACK

Drawn by Captain May.

The Voyage of the 'Fox' in the Arctic Seas.

A NARRATIVE
OF THE
DISCOVERY OF THE FATE
OF
SIR JOHN FRANKLIN
AND
HIS COMPANIONS.

BY CAPTAIN M'CLINTOCK, R.N., LL.D.

With Maps and Illustrations.

LONDON:
JOHN MURRAY, ALBEMARLE STREET,
PUBLISHER TO THE ADMIRALTY.
1859.

DEDICATION.

My dear Lady Franklin,

There is no one to whom I could with so much propriety or willingness dedicate my Journal as to you. For you it was originally written, and to please you it now appears in print.

To our mutual friend, Sherard Osborn, I am greatly obliged for his kindness in seeing it through the press—a labour I could not have settled down to so soon after my return; and also for pointing out some omissions and technicalities which would have rendered parts of it unintelligible to an ordinary reader. These kind hints have been but partially attended to, and, as time presses, it appears with the mass of its original imperfections, as when you read it in manuscript. Such as it is, however, it affords me this valued opportunity of assuring you of the real gratification I feel in having been instrumental in accomplishing an object so dear to you. To your devotion and self-sacrifice the world is indebted for the deeply-interesting revelation unfolded by the voyage of the 'Fox.'

Believe me to be,

With sincere respect, most faithfully yours,

F. L. M'CLINTOCK.

London, 24*th* *Nov.* 1859.

LIST OF OFFICERS AND SHIP'S COMPANY OF THE 'FOX.'

Name	Rank	Remarks
F. L. M'Clintock ..	Captain R.N.	
W. R. Hobson	Lieutenant R.N.	
Allen W. Young	Captain, Mercantile Marine.	
David Walker, M.D.	Surgeon and Naturalist.	
George Brands	Engineer	Died 6th Nov. 1858 (Apoplexy).
Carl Petersen	Interpreter.	
Thomas Blackwell ..	Ship's Steward	Died 14th June, 1859 (Scurvy).
Wm. Harvey	Chief Quartermaster.	
Henry Toms	Quartermaster.	
Alex. Thompson	,,	
John Simmonds	Boatswain's Mate.	
George Edwards	Carpenter's Mate.	
Robert Scott	Leading Stoker	Died 4th Dec. 1857 (in consequence of a fall).
Thomas Grinstead ..	Sailmaker.	
George Hobday	Captain of Hold.	
Robert Hampton	A. B.	
John A. Haselton ..	,,	
George Carey	,,	
Ben. Pound	,,	
Wm. Walters	Carpenter's Crew.	
Wm. Jones	Dog-driver.	
James Pitcher	Stokers.	
Thomas Florance ..		
Richard Shingleton ..	Officers' Steward.	
Anton Christian	Greenland Esquimaux	Discharged in Greenland.
Samuel Emanuel		

OFFICIAL ACKNOWLEDGMENT OF THE SERVICES OF THE YACHT 'FOX.'

ADMIRALTY, LONDON,
24*th Oct.* 1859.

SIR,

I am commanded by my Lords Commissioners of the Admiralty to acquaint you, that, in consideration of the important services performed by you in bringing home the only authentic intelligence of the death of the late Sir John Franklin, and of the fate of the crews of the 'Erebus' and 'Terror,' Her Majesty has been pleased, by her order in Council of the 22nd instant, to sanction the time during which you were absent on these discoveries in the Arctic Regions, viz. from the 30th June 1857 to the 21st September 1859, to reckon as time served by a captain in command of one of Her Majesty's ships, and my Lords have given the necessary directions accordingly.

I am, Sir,

Your very humble servant,

W. G. ROMAINE,

Secretary to the Admiralty.

Captain Francis L. M'Clintock, R.N.

PREFACE.

The following narrative of the bold adventure which has successfully revealed the last discoveries and the fate of Franklin, is published at the request of the friends of that illustrious navigator. The gallant M'Clintock, when he penned his journal amid the Arctic ices, had no idea whatever of publishing it; and yet there can be no doubt that the reader will peruse with the deepest interest the simple tale of how, in a little vessel of 170 tons burthen, he and his well-chosen companions have cleared up this great mystery.

To the honour of the British nation, and also let it be said to that of the United States of America, many have been the efforts made to discover the route followed by our missing explorers. The highly deserving men who have so zealously searched the Arctic seas and lands in this cause must now rejoice, that after all their anxious toils, the merit of rescuing from the frozen North the record of the last days of Franklin, has fallen to the share of his noble-minded widow.

Lady Franklin has, indeed, well shown what a devoted and true-hearted Englishwoman can accomplish. The moment that relics of the expedition commanded by her husband were brought home (in 1854) by Rae, and that she heard of the account given to him by the Esquimaux of a large party of Englishmen having been seen struggling with difficulties on the ice near the mouth of the Back or Great Fish River, she resolved to expend all her available means (already much exhausted in four other independent expeditions) in an exploration of the limited area to which the search must thenceforward be necessarily restricted.

Whilst the supporters of Lady Franklin's efforts were of opinion, that the Government ought to have undertaken a search, the extent of which was, for the first time, definitely limited, it is but rendering justice to the then Prime Minister* to state, that he had every desire to carry out the wishes of the men of science† who appealed to him, and that he was

* Viscount Palmerston.

† See the Memorial (Appendix) addressed to the First Lord of the Treasury, headed by Admiral Sir F. Beaufort, General Sabine, and many other men of science, and which, as President of the Royal

precluded from acceding to their petition, by nothing but the strongly expressed opinion of official authorities, that after so many failures the Government were no longer justified in sending out more brave men to encounter fresh dangers in a cause which was viewed as hopeless. Hence it devolved on Lady Franklin and her friends to be the sole means of endeavouring to bring to light the true history of her husband's voyage and fate.

Looking to the list of Naval worthies who during the preceding years had been exploring the Arctic Regions, Lady Franklin was highly gratified when she obtained the willing services of Captain M'Clintock to command the yacht 'Fox,' which she had purchased; for that officer had signally distinguished himself in the voyages of Sir James Ross and Captain (now Admiral) Austin, and especially in his extensive journeys on the ice when associated with Captain Kellett. With such a leader she could not but entertain sanguine hopes of success when the fast and well-adapted little vessel

---

Geographical Society, I presented to the Prime Minister; and also the speech of Lord Wrottesley, the President of the Royal Society, who, in the absence of the lamented Earl of Ellesmere, brought the subject earnestly under the notice of the House of Lords on the 18th of July, 1856.

sailed from Aberdeen on the 1st of July, 1857, upon this eventful enterprise.

Deep, indeed, was the mortification experienced by every one who shared the feelings and anticipations of Lady Franklin when the untoward news came, in the summer of 1858, that, the preceding winter having set in earlier than usual, the 'Fox' had been beset in the ice off Melville Bay, on the coast of Greenland, and after a dreary winter, various narrow escapes, and eight months of imprisonment, had been carried back by the floating ice nearly twelve hundred geographical miles—even to 63½° N. lat. in the Atlantic! See the woodcut map, No. 1.

But although the good little yacht had been most roughly handled among the ice-floes (see Frontispiece), we were cheered up by the information from Disco, that, with the exception of the death of the engine-driver in consequence of a fall into the hold, the crew were in stout health and full of energy, and that, provided with sufficient fuel and provisions, a good supply of sledging dogs, two tried Esquimaux, and the excellent interpreter Petersen the Dane,* ample grounds yet remained to

* Since his return to Copenhagen, Petersen has been worthily honoured by his Sovereign with the silver cross of Dannebrog.

FIG. 1.

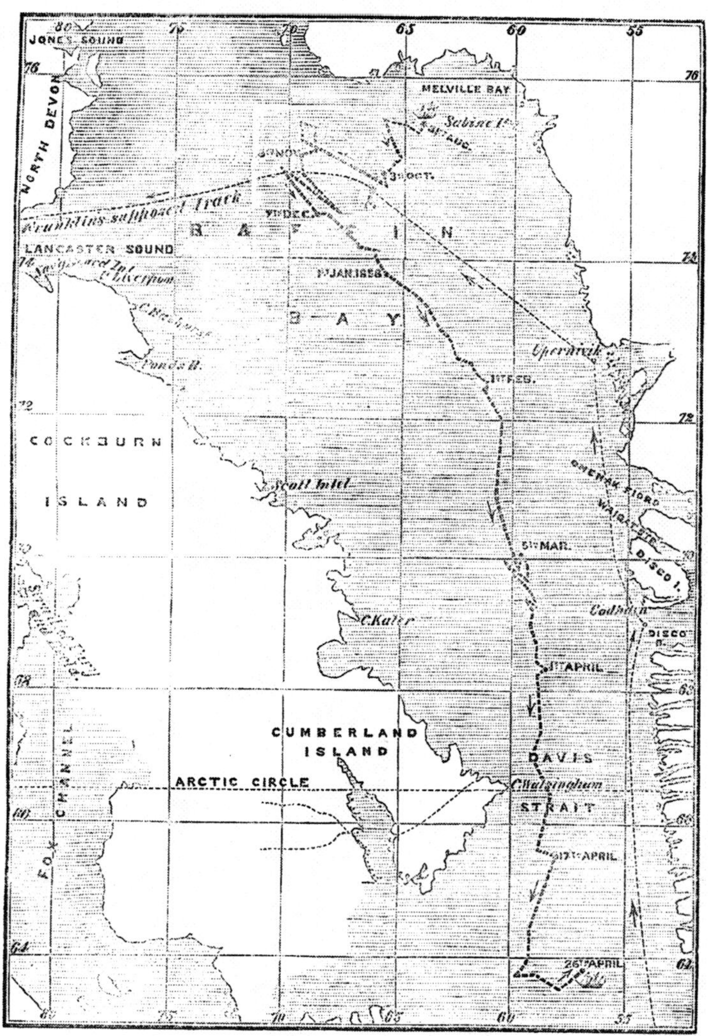

SKETCH MAP OF THE DRIFT OF THE 'FOX' DOWN BAFFIN'S BAY IN THE FLOATING ICE.

FIG. 2.

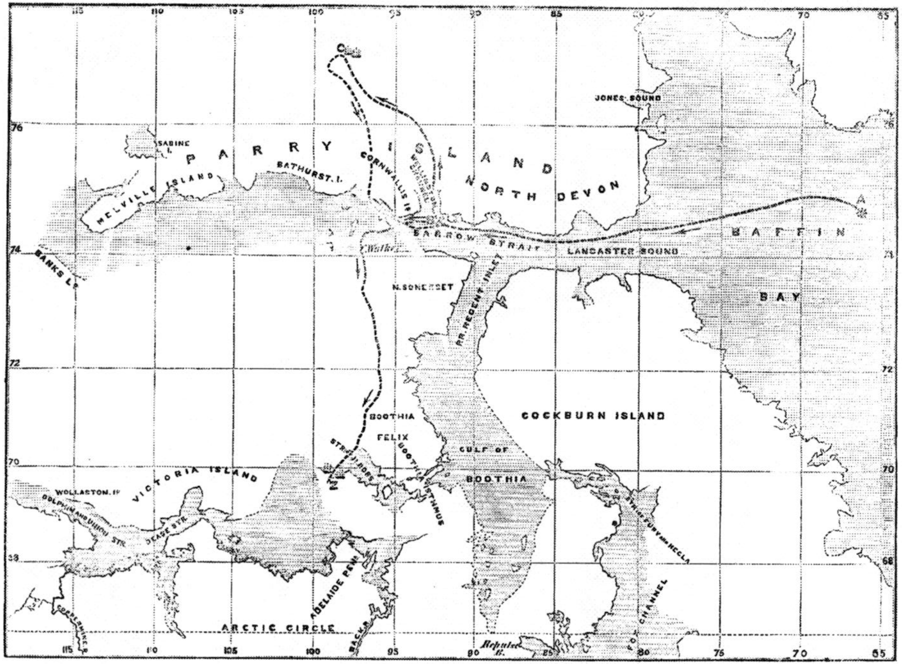

SKETCH MAP OF ARCTIC REGIONS AT THE TIME OF FRANKLIN'S LAST EXPEDITION.

lead us to hope for a successful issue. Above all, we were encouraged by the proofs of the self-possession and calm resolve of M'Clintock, who held steadily to the accomplishment of his original project; the more so as he had then tested and recognised the value of the services of Lieutenant (now Commander) Hobson, his able second in command; of Captain Allen Young, his generous volunteer associate;* and of Dr. Walker, his accomplished Surgeon.

Despite, however, of these reassuring data, many an advocate of this search was anxiously alive to the chance of the failure of the venture of one unassisted yacht, which after sundry mishaps was again starting to cross Baffin's Bay, with the foreknowledge, that when she reached the opposite coast, the real difficulties of the enterprise were to commence.

Any such misgivings were happily illusory; and the reader who follows M'Clintock across the "middle ice" of Baffin's Bay to Pond Inlet, thence to Beechey Island, down a portion of Peel Strait, and then through the hitherto unnavigated waters of Bellot Strait in one summer

* Captain Allen Young of the merchant marine not only threw his services into this cause, and subscribed 500*l*. in furtherance of the expedition, but, abandoning lucrative appointments in command, generously accepted a subordinate post.

season, may reasonably expect the success which followed.

Whilst the revelation obtained from the long-sought records, which were discovered by Lieutenant Hobson, is most satisfactory to those who speculated on the probability of Franklin having, in the first instance, tried to force his way northwards through Wellington Channel (as we now learn he did), those who held a different hypothesis, namely, that he followed his instructions, which directed him to the S.-W., may be amply satisfied, that in the following season the ships did pursue this southerly course till they were finally beset in N. lat. 70° 05′.*

At the same time, the public should fully understand the motive which prompted the supporters of Lady Franklin in advocating this last search. Putting aside the hope which some of us entertained, that a few of the younger men of the missing expedition might still be found to be

* For a *résumé* of all the plans of research and the speculations of seamen and geographers, see the interesting and most useful volume of Mr. John Brown, entitled, 'The North-West Passage and Search after Sir John Franklin,' 1858. In an Appendix to this work we learn, that from the earliest Polar researches by John Cabot, at the end of the 15th century, to the voyage of M'Clintock, there have been about 130 expeditions, illustrated by 250 books and printed documents, of which 150 have been issued in England. Amidst the various recent publications, it is but rendering justice to Dr. King, the former companion of Sir George Back, to state that he suggested and always maintained the necessity of a search for the missing navigators at or near the mouth of the Back River.

living among the Esquimaux, we had every reason to expect, that if the ships were discovered, the scientific documents of the voyage, including valuable magnetic observations, would be recovered.

In the absence of such good fortune we may, however, well be gladdened by the discovery of that one precious document which gives us a true outline of the voyage of the 'Erebus' and 'Terror.'

That the reader may comprehend the vast extent of sea traversed by Franklin in the two summers before his ships were beset, a small map (No. 2) is here introduced representing all the lands and seas of the Arctic regions to the west of Lancaster Sound which were known and laid down when he sailed. The dotted lines and arrows, which extend from the then known seas and lands into the unknown waters or blank spaces on this old map indicate Franklin's route, the novelty, range, rapidity, and boldness of which, as thus delineated, may well surprise the geographer, and even the most enterprising Arctic sailor.* For, those who

* The letter A in Baffin Bay (fig. 1) indicates the spot where Franklin was last seen. In fig. 2, B is the winter rendezvous at Beechey Island; C, the greatest northing of the expedition, viz. 77° N. lat.; Z, the final beset of the 'Erebus' and 'Terror;' the extreme north and south points of their voyage being represented by two small ships.

have not closely attended to the results of other Arctic voyages may be informed, that rarely has an expedition in the first year accomplished more by its ships, than the establishing of good winter quarters, from whence the real researches began by sledge-work in the ensuing spring. Franklin, however, not only reached Beechey Island, but ascended Wellington Channel, then an unknown sea, to 77° N. lat., a more northern latitude in this meridian than that attained long afterwards in ships by Sir Edward Belcher, and much to the north of the points reached by Penny and De Haven. Next, though most scantily provided with steam-power, Franklin navigated round Cornwallis' Land, which he thus proved to be an island. This last discovery of a navigable channel throughout, between Cornwallis and Bathurst Islands, though made in the very summer he left England, has remained even to this day unknown to other navigators!

Franklin then, in obedience to his orders, steered to the south-west. Passing, as M'Clintock believes, down Peel's Strait in 1846, and reaching as far as lat. 70° 05′ N., and long. 98° 23′ W., where the ships were beset, it is clear that he, who, with others, had previously ascertained the existence of a channel along the north coast of America, with which the sea wherein he was interred had a direct communication, was the *first real discoverer of the North-West Passage*. This great fact must therefore be inscribed upon the monument of Franklin.

The adventurous M'Clure, who has been worthily honoured for working out another North-Western passage, which we now know to have been of subsequent date,* as well as Collinson, who, taking the 'Enterprise' along the north coast of America, and afterwards bringing her home, reached with sledges the western edge of the area recently laid open by M'Clintock, will I have no doubt unite with their Arctic associates, Richardson, Sherard Osborn, and M'Clintock, in affirming, that "Franklin and his followers secured the honour for which they died — that of being the first discoverers of the North-West Passage." †

* In 1850.

† See a most heart-stirring sketch of the last voyage of Sir John Franklin by Captain Sherard Osborn, in the periodical *Once a Week*, of the 22nd and 29th October and 5th November last. Possessing a thorough acquaintance with the Arctic regions, this distinguished seaman has shown more than his ordinary power of description, in placing before the public his conception of what may have been the chief occurrences in the voyage of the 'Erebus' and 'Terror,' and the last days of Franklin, as founded upon an acquaintance with the character of the chief and his associates, and the record and relics obtained by M'Clintock. This sketch is prefaced by a spirited and graceful outline of all previous geographical discoveries, from the day when they were originated by the father

Again, when we turn from the discoveries of Franklin to those of M'Clintock, as mapped in red colours on the general map, on which is represented the amount of outline laid down by all other Arctic explorers from the days when these modern researches originated with Sir John Barrow, we perceive that, in addition to the discovery of the course followed by the 'Erebus' and 'Terror,' some most important geographical data have been accumulated by the last expedition of Lady Franklin.

Thus, M'Clintock has proved, that the strait named by Kenedy in an earlier private expedition of Lady Franklin after his companion the brave Lieutenant Bellot, and which has hitherto been regarded only as an impassable frozen channel, or ignored as a channel at all, is a navigable strait, the south shore of which is thus seen to be the northernmost land of the continent of America.

M'Clintock has also laid down the hitherto unknown coast-line of Boothia, southwards from Bellot Strait to the Magnetic Pole, has delineated the whole of King William's Island, and

---

of all modern Arctic enterprise, Sir John Barrow, to whom, and to many other eminent persons, from Sir Edward Parry downwards, I have in various Geographical Addresses offered the tribute of my admiration.

opened a new and capacious, though ice-choked channel, suspected before, but not proved, to exist, extending from Victoria Strait in a north-west direction to Melville or Parry Sound. The latter discovery rewarded the individual exertions of Captain Allen Young, but will very properly, at Lady Franklin's request, bear the name of the leader of the 'Fox' expedition, who had himself assigned to it the name of the widow of Franklin.*

Neither has the expedition been unproductive of scientific results. For, whilst many persons will be interested in the popular descriptions of the native Esquimaux, as well as of the lower animals, the man of science will hereafter be further gratified by having presented to him, in the form of an additional Appendix,† most valuable details relating to the zoology, botany, meteorology, and especially to the terrestrial magnetism, of the region examined.

Lastly, M'Clintock has convinced himself, that the best way of securing the passage of a ship

---

* In his volume before cited, p. xii., Mr. John Brown gave strong reasons (which he had held for some time) for believing in the existence of the very channel which now bears the name of M'Clintock. It is, however, the opinion both of that officer and his associates, as also of Captain Sherard Osborn, that Franklin could not have reached the spot where his ships were beset by proceeding down that ice-choked channel, but that he must have sailed down Peel Sound.

† Much of this Appendix will be prepared by Dr. David Walker.

from the Atlantic to the Pacific, is by following, as near as possible, the coast-line of North America: indeed, it is his opinion, founded upon a large experience, that no passage by a ship can ever be accomplished in a more northern direction. This it is well known was the favourite theory of Franklin, who had himself, along with Richardson, Back, Beechey, Dease, Simpson, and Rae, surveyed the whole of that same North American coast from the Back or Great Fish River to Behring Strait. Thus, when Franklin sailed in 1845, the discovery of a North-West Passage was reduced to the finding a link between the latter survey and the discoveries of Parry, who had already, to his great renown, opened the first half of a more northern course from east to west, when he was arrested by the impenetrable ice-barrier at Melville Island.

And here it is to be remembered, that the tract in which the record and the relics have been found, is just that to which Lady Franklin herself specially directed Kenedy, the commander of the 'Prince Albert,' in her second private expedition in 1852; and had that intrepid explorer not been induced to search northwards of Bellot Strait, but had felt himself able to follow the course indicated by his sagacious employer, there can be no doubt, that much more satisfactory results would have been obtained than those which, after a lapse of seven years, have now been realized by the undaunted perseverance of Lady Franklin, and the skill and courage of M'Clintock.

The natural modesty of this commander has, I am bound to say, prevented his doing common justice, in the following journal, to his own conduct—conduct which can be estimated by those only who have listened to the testimony of the officers serving with and under the man, whose great qualities in moments of extreme peril elicited their heartiest admiration and ensured their perfect confidence.

In writing this Preface (which I do at the request of the promoters of the last search), I may state that, having occupied the Chair of the Royal Geographical Society in 1845, when my cherished friend, Sir John Franklin, went forth for the third time to seek a North-West passage, it became my bounden duty in subsequent years, when his absence created much anxiety, and when I re-occupied the same position, ardently to promote the employment of searching expeditions, and warmly to sustain Lady Franklin's endeavours in this holy cause.

Imbued with such feelings, I must be permitted to say, that no event in my life gave me purer delight, than when Captain Collinson, whose labours to support and carry out this last search have been signally serviceable, forwarded to me a telegram to be communicated to the British Association at Aberdeen announcing the success of M'Clintock. That document reached Balmoral on the 22nd of September last, when the men of science were invited thither by their Sovereign. Great was the satisfaction caused by the diffusion of these good tidings among my associates (the distinguished Arctic explorers Admiral Sir James Ross and General Sabine being present); and it was most cheering to us to know, that the Queen and our Royal President* took the deepest interest in this intelligence — such as, indeed, they have always evinced whenever the search for the missing navigators has been brought under their consideration. The immediate bestowal of the

---

* At the Aberdeen meeting the Prince Consort thus spoke :— "The Aberdeen whaler braves the icy regions of the Polar sea to seek and to battle with the great monster of the deep; he has materially assisted in opening these icebound regions to the researches of science; he fearlessly aided in the search after Sir John Franklin and his gallant companions whom their country sent forth on this mission; but to whom Providence, alas! has denied the reward of their labours, the return to their homes, to the affectionate embrace of their families and friends, and the acknowledgments of a grateful nation."

Arctic medal upon all the officers and men of the 'Fox' is a pleasing proof that this interest is well sustained.

But these few introductory sentences must not be extended; and I invite the reader at once to peruse the Journal of M'Clintock, which will gratify every lover of truthful and ardent research, though it will leave him impressed with the sad belief, that the end of the companions of Franklin has been truly recorded by the native Esquimaux, who saw these noble fellows "fall down and die as they walked along the ice."

Looking to the fact, that little or no fresh food could have been obtained by the crews of the 'Erebus' and 'Terror' during their long imprisonment of twenty months, in so frightfully sterile a region as that in which the ships were abandoned,—so sterile that it is even deserted by the Esquimaux,—and also to the want of sustenance in spring at the mouth of the Back River, all the Arctic naval authorities with whom I have conversed, coincide with M'Clintock and his associates in the belief, that none of the missing navigators can be now living.

Painful as is the realisation of this tragic event, let us now dwell only on the reflection that, while the North-West passage has been solved by the heroic self-sacrifice of Franklin, Crozier, Fitzjames, and their associates, the searches after them which are now terminated, have, at a very small loss of life, not only added prodigiously to geographical knowledge, but have, in times of peace, been the best school for testing, by the severest trials, the skill and endurance of many a brave seaman. In her hour of need—should need arise—England knows that such men will nobly do their duty.

RODERICK I. MURCHISON.

CONTENTS.

CHAPTER VI.

CHAPTER VII.

CHAPTER VIII.

CHAPTER IX.

CHAPTER X.

CHAPTER XI.

CHAPTER XII.

CHAPTER XIII.

CHAPTER XIV.

CHAPTER XV.

CHAPTER XVI.

CHAPTER XVII.

APPENDIX.

LIST OF ILLUSTRATIONS.

JOURNAL OF THE SEARCH

FOR

SIR JOHN FRANKLIN.

CHAPTER I.

Cause of delay in equipment — Fittings of the 'Fox' — Volunteers for Arctic service — Assistance from public departments — Reflections upon the undertaking — Instructions and departure — Orkneys and Greenland — Fine Arctic scenery — Danish establishments in Greenland — Fredrickshaab, in Davis' Straits.

It is now a matter of history how Government and private expeditions prosecuted with unprecedented zeal and perseverance the search for Sir John Franklin's ships, between the years 1847–55; and that the only ray of information gleaned was that afforded by the inscriptions upon three tombstones at Beechey Island, briefly recording the names and dates of the deaths of those individuals of the lost expedition, who thus early fell in the cause of science and of their country.

In this manner were we made aware of the

locality where the Franklin expedition passed its first arctic winter. The traces assuring us of that fact were discovered in August, 1850, by Captain Ommanney, R.N., of H.M.S. 'Assistance,' and by Captain Penny of the 'Lady Franklin.'

In October, 1854, Dr. Rae brought home the only additional information respecting them which has ever reached us. From the Esquimaux of Boothia Felix he learned that a party of about forty white men were met on the west coast of King William's Island, and from thence travelled on to the mouth of the Great Fish River, where they all perished of starvation, and that this tragic event occurred apparently in the spring of 1850.

Some relics obtained from these natives, and brought home by Dr. Rae, were proved to have belonged to Sir John Franklin and several of his associates.

The Government caused an exploring party to descend the Fish River in 1855; but, although sufficient traces were found to prove that some portion of the crews of the 'Erebus' and 'Terror' had actually landed on the banks of that river, and traces existed of them up to Franklin Rapids, no additional information was obtained either from the discovery of records, or through the Esquimaux. Mr. Anderson, the Hudson Bay Company's officer in charge, and his small party, deserve credit for their perseverance and skill; but they were not furnished with the necessary means of accomplishing their mission. Mr. Anderson could not obtain an interpreter, and the two frail bark canoes in which his whole party embarked were almost worn out before they reached the locality to be searched. It is not surprising that such an expedition caused very considerable disappointment at home.

Lady Franklin, and the advocates for further search, now pressed upon Government the necessity of following up, in a more effectual manner, the traces accidentally found by Dr. Rae, and, in fact, of rendering the search complete by one more effort, involving but little of hazard or expense. It was not until April, 1857, that any decisive answer was given to Lady Franklin's appeal. (See Appendix No. 1.)

Sir Charles Wood then stated "that the members of Her Majesty's Government, having come, with great regret, to the conclusion that there was no prospect of saving life, would not be justified, for any objects which in their opinion could be obtained by an expedition to the Arctic seas, in exposing the lives of officers and men to the risk inseparable from such an enterprise."

Lady Franklin, upon this final disappointment of her hopes, had no hesitation in immediately preparing to send out a searching expedition, equipped and stored at her own cost. But she was not left alone. Many friends of the cause—including some of the most distinguished scientific men in England, and especially Sir Roderick Murchison, whose zeal was as practical as it was enlightened—hastened to tender their aid, and soon a very considerable sum was raised in furtherance of so truly noble an effort.

On the 18th April, 1857, Lady Franklin did me the honour to offer me the command of the proposed expedition,—it was of course most cheerfully accepted. As a post of honour and of some difficulty it possessed quite sufficient charms for a naval officer who had already served in three consecutive expeditions from 1848 to 1854. I was thoroughly conversant with all the details of this peculiar service; and I confess, moreover, that my whole heart was in the cause. How could I do otherwise than devote myself to save at least the record of faithful service, even unto death, of my brother officers and seamen? and, being one of those by whose united efforts not only the Franklin search, but the geography of Arctic America, has been brought so nearly to completion, I could not willingly resign to posterity, the honour of filling up even the small remaining blank upon our maps.

To leave these discoveries incomplete, more especially in a quarter through which the tidal stream actually demonstrates the existence of a channel—the only remaining hope of a practicable north-west passage — would indeed be leaving strong inducement for future explorers to reap the rich reward of our long-continued exertions.

I immediately applied to the Admiralty for leave of absence to complete the Franklin search; and on the 23rd received at Dublin the telegraphic message from Lady Franklin: "Your leave is granted; the 'Fox' is mine; the refit will commence immediately." She had already purchased the screw-yacht 'Fox,' of 177 tons burthen, and now placed her, together with the necessary funds, at my disposal.

Let me explain what is here implied by the simple word refit. The velvet hangings and splendid furniture of the yacht, and also everything not constituting a part of the vessel's strengthening, were to be removed; the large skylights and capacious ladderways had to be reduced to limits more adapted to a polar clime; the whole vessel to be externally sheathed with stout planking, and internally fortified by strong cross beams, longitudinal beams, iron stanchions, and diagonal fastenings; the false keel taken off, the slender brass propeller replaced by a massive iron one, the boiler taken out, altered, and enlarged; the sharp stem to be cased in iron until it resembled a ponderous chisel set up edgeways; even the yacht's rig had to be altered.

She was placed in the hands of her builders, Messrs. Hall and Co., of Aberdeen, who displayed even more than their usual activity in effecting these necessary alterations, for it was determined that the 'Fox' should sail by the 1st July.

Internally she was fitted up with the strictest economy in every sense, and the officers were crammed into pigeon-holes, styled cabins, in order to make room for provisions and stores; our mess-room, for five persons, measured 8 feet square. The ordinary heating apparatus for winter use was dispensed with, and its place supplied by a few very small stoves. The 'Fox' had been the property of the late Sir Richard Sutton, Bart., who made but one trip to Norway in her, and she was purchased by Lady Franklin from his executors for 2000*l.*

Having thus far commenced the refit of the vessel, I turned my attention to the selection of a crew and to the requisite clothing and provisions for our voyage.

Many worthy old shipmates, my companions in the previous Arctic voyages, most readily volunteered their services, and they were as cheerfully accepted, for it was my anxious wish to gather around me well-tried men, who were aware of the duties expected of them, and accustomed to naval discipline. Hence, out of the twenty-five souls composing our small company, seventeen had previously served in the Arctic search.

Expeditions of this nature are always popular with seamen, and innumerable were the applications sent to me; but still more abundant were the offers to "serve in any capacity" which poured in from all parts of the country, from people of all classes, many of whom had never seen the sea. It was, of course, impossible to accede to any of these latter proposals, yet, for my own part, I could not but feel gratified at such convincing proofs that the spirit of the country was favourable to us, and that the ardent love of hardy enterprise still lives amongst

Englishmen, as of old, to be cherished, I trust, as the most valuable of our national characteristics —as that which has so largely contributed to make England what she is.

My second in command was Lieutenant W. R. Hobson, R.N., an officer already distinguished in Arctic service. Captain Allen Young joined me as sailing-master, contributing not only his valuable services but largely of his private funds to the expedition. This gentleman had previously commanded some of our very finest merchant ships, the latest being the steam-transport 'Adelaide' of 2500 tons: he had but recently returned, in ill-health, from the Black Sea, where he was most actively employed during the greater part of the Crimean campaign. Nothing that I could say would add to the merit of such singularly generous and disinterested conduct. David Walker, M.D., volunteered for the post of surgeon and naturalist; he also undertook the photographic department; and just before sailing, Carl Petersen, now so well known to Arctic readers as the Esquimaux interpreter in the expeditions of Captain Penny and Dr. Kane, came to join me from Copenhagen, although landed there from Greenland only six days previously, after an absence of a year from his family: we were indebted to Sir

Roderick Murchison and the electric telegraph for securing his valuable services.

Like the Paris omnibuses we were at length *tout complet*, and quite as anxious to make a start.

Ample provisions for twenty-eight months were embarked, including preserved vegetables, lemon-juice, and pickles, for daily consumption, and preserved meats for every third day: also as much of Messrs. Allsopp's stoutest ale as we could find room for. The Government, although declining to send out an expedition, yet now contributed liberally to our supplies. All our arms, powder, shot, powder for ice-blasting, rockets, maroons, and signal-mortar, were furnished by the Board of Ordnance. The Admiralty caused 6682 lbs. of pemmican to be prepared for our use. Not less than 85,000 lbs. of this invaluable food have been prepared since 1845 at the Royal Clarence Victualling Yard, Gosport, for the use of the Arctic Expeditions. It is composed of prime beef cut into thin slices and dried over a wood fire; then pounded up and mixed with about an equal weight of melted beef fat. The pemmican is then pressed into cases capable of containing 42 lbs. each. The Admiralty supplied us also with all the requisite ice-gear, such as saws from ten to eighteen feet in length, ice-anchors, and ice-claws: also with our winter housing, medicines, pure lemon-juice, seamen's library, hydrographical instruments, charts, chronometers, and an ample supply of arctic clothing which had remained in store from former expeditions. The Board of Trade contributed a variety of meteorological and nautical instruments and journals; and I found that I had but to ask of these departments for what was required, and if in store it was at once granted. I asked, however, only for such things as were indispensably necessary.

The President and Council of the Royal Society voted the sum of *50l.* from their donation fund for the purchase of magnetic and other scientific instruments, in order that our anticipated approach to so interesting a locality as the Magnetic Pole might not be altogether barren of results.

Being desirous to retain for my vessel the privileges she formerly enjoyed as a yacht, my wishes were very promptly gratified; in the first instance by the Royal Harwich Yacht Club, of which my officers and myself were enrolled as members—the Commodore, A. Arcedeckne, Esq., presenting my vessel with the handsome ensign and burgee of the Club; and shortly afterwards by my being elected a member of the Royal Victoria Yacht Club for the period of my voyage. Lastly, upon the very day of sailing, I was proposed for the Royal Yacht Squadron, to which the yacht had previously belonged when the property of Sir Richard Sutton.

Throughout the whole period required for our equipment I constantly experienced the heartiest co-operation and earnest goodwill from all with whom my varied duties brought me in contact. Deep sympathy with Lady Franklin in her distress, her self-devotion and sacrifice of fortune, and an earnest desire to extend succour to any chance survivors of the ill-fated expedition who might still exist, or, at least, to ascertain their fate, and rescue from oblivion their heroic deeds, seemed the natural promptings of every honest English heart. It is needless to add that this experience of public opinion confirmed my own impression that the glorious mission intrusted to me was in reality a *great national duty*. I could not but feel that, if the gigantic and admirably equipped national expeditions sent out upon precisely the same duty, and reflecting so much credit upon the Board of Admiralty, were ranked amongst the noblest efforts in the cause of humanity any nation ever engaged in, and that, if high honour was awarded to all composing those splendid expeditions, surely the effort became still more remarkable and worthy of approbation when its means were limited to one little vessel, containing but twenty-five souls, equipped and provisioned (although efficiently, yet) in a manner more according with the limited resources of a private individual than with those of the public purse. The less the means, the more arduous I felt was the achievement. The greater the risk—for the 'Fox' was to be launched alone into those turbulent seas from which every other vessel had long since been withdrawn—the more glorious would be the success, the more honourable even the defeat, if again defeat awaited us.

Upon the last day of June Lady Franklin, accompanied by her niece Miss Sophia Cracroft, and Capt. Maguire, R.N., came on board to bid us farewell, for we purposed sailing in the evening. Seeing how deeply agitated she was on leaving the ship, I endeavoured to repress the enthusiasm of my crew, but without avail; it found vent in three prolonged hearty cheers. The strong feeling which prompted them was truly sincere; and this unbidden exhibition of it can hardly have gratified her for whom it was intended more than it did myself.

I must here insert the only written instructions I could prevail upon Lady Franklin to give me; they were not read until the 'Fox' was fairly in the Atlantic.

Aberdeen, June 29, 1857.

MY DEAR CAPTAIN M'CLINTOCK,

You have kindly invited me to give you "Instructions," but I cannot bring myself to feel that it would be right in me in any way to influence your judgment in the conduct of your noble undertaking; and indeed I have no temptation to do so, since it appears to me that your views are almost identical with those which I had independently formed before I had the advantage of being thoroughly possessed of yours. But had this been otherwise, I trust you would have found me ready to prove the implicit confidence I place in you by yielding my own views to your more enlightened judgment; knowing too as I do that your whole heart also is in the cause, even as my own is. As to the objects of the expedition and their relative importance, I am sure you know that the rescue of any possible survivor of the 'Erebus' and 'Terror' would be to me, as it would be to you, the noblest result of our efforts.

To this object I wish every other to be subordinate; and next to it in importance is the recovery of the unspeakably precious documents of the expedition, public and private, and the personal relics of my dear husband and his companions.

And lastly, I trust it may be in your power to confirm, directly or inferentially, the claims of my husband's expedition to the earliest discovery of the passage, which, if Dr. Rae's report be true (and the Government of our country has accepted and rewarded it as such), these martyrs in a noble cause achieved at their last extremity, after five long years of labour and suffering, if not at an earlier period.

I am sure you will do all that man can do for the attainment of all these objects; my only fear is that you may spend yourselves too much in the effort; and you must therefore let me tell you how much dearer to me even than any of them is the preservation of the valuable lives of the little band of heroes who are your companions and followers.

May God in his great mercy preserve you all from harm amidst the labours and perils which await you, and restore you to us in health and safety as well as honour! As to the honour I can have *no* misgiving. It will be yours as much if you fail (since you *may* fail in spite of every effort) as if you succeed; and be assured that, under *any and all circumstances whatever*, such is my unbounded confidence in you, you will possess and be entitled to the enduring gratitude of your sincere and attached friend,

JANE FRANKLIN.

We were not destined to get to sea that evening. The 'Fox,' hitherto during her brief career, accustomed only to the restraint imposed upon a gilded pet in summer seas, seemed to have got an inkling that her duty henceforth was to combat with difficulties, and, entering fully into the spirit of the cruize, answered her helm so much more readily than the pilot expected that she ran aground upon the bar. She was promptly shored up, and remained in that position until next morning, when she floated off unhurt at high water, and commenced her long and lonely voyage.

Scarcely had we left the busy world behind us when we were actively engaged in making arrangements for present comfort and future exertion. How busy, how happy, and how full of hope we all were then!

On the night of the 2nd of July we passed through the Pentland Firth, where the tide rushing impetuously against a strong wind raised up a tremendous sea, amid which the little vessel struggled bravely under steam and canvas. The bleak wild shores of Orkney, the still wilder pilot's crew, and their hoarse screams and unintelligible dialect, the shrill cry of innumerable sea-birds, the howling breeze and angry sea, made us feel as if we had suddenly awoke in Greenland itself. The southern extremity of that ice-locked continent became visible on the 12th. It is quaintly named Cape Farewell; but whether by some sanguine outward-bound adventurer who fancied that in leaving Greenland behind him he had already secured his passage to Cathay; or whether by the wearied homesick mariner, feebly escaping from the grasp of winter in his shattered bark, and firmly purposing to bid a long farewell to this cheerless land, history altogether fails to enlighten us.

From January until July this coast is usually rendered unapproachable by a broad margin of heavy ice, which drifts there from the vicinity of Spitzbergen, and, lapping round the Cape, extends alongshore to the northward about as far as Baal's River, a distance of 250 miles. Although it effectually blockades the ports of South Greenland for the greater part of the summer, and is justly dreaded by the captains of the Greenland traders, it confers important benefits upon the Greenlander by bearing to his shores immense numbers of seals and many bears. The same current which conveys hither all this ice is also freighted with a scarcely less valuable supply of driftwood from the Siberian rivers.

About this time, one of my crew showing symptoms of diseased lungs, I determined to embrace the earliest opportunity of sending him home out of a climate so fatal to those who are thus affected; and having learnt from Mr. Petersen, who had quitted Greenland only in April last, that a vessel would very soon leave Frederickshaab for Copenhagen, I resolved to go to that place in order to catch this homeward-bound ship.

It was necessary to push through the Spitzbergen ice, and we fortunately succeeded in doing so after eighteen hours of buffeting with this formidable enemy; at first we found it tolerably loose, and the wind being strong and favourable, we thumped along pleasantly enough; but as we advanced, the ice became much more closely packed, a thick fog came on, and many hard knocks were exchanged; at length our steam carried us through into the broad belt of clear water between the ice and land, which Petersen assured me always exists here at this season.

The dense fog now prevented further progress, and as evening closed in I gave up all hope of improvement for the night, when suddenly the fog rolled back upon the land, disclosing some islets close to us, then the rugged points of mainland, and at length, lifting altogether, the distant snowy mountain-peaks against a deep blue sky.

The evening became bright and delightful; the whole extent of coast was fringed with innumerable islets, backed by lofty mountains, and, being richly tinted by a glorious western sun, formed an unusually splendid sight. Greenland unveiled to our anxious gaze that memorable evening, all the magnificence of her natural beauty. Was it to welcome us that she thus cast off her dingy outer mantle and shone forth radiant with smiles?—such winning smiles!

A faint streak of mist, which we could not account for, appeared to float across a low wide interval in the mountain range; the telescope revealed its true character,—it was a portion of the distant glacier. We found ourselves upon the Tallard Bank, 30 miles north of our port, having been rapidly carried northwards by the Spitzbergen current.

July 20th.—This morning the chief trader of the settlement, or, as he is more usually styled by the English, the Governor, came off to us, and his pilot soon conducted us into the safe little harbour of Frederickshaab. I was much gratified to learn that we were just in time to secure a passage home for our ailing shipmate.

For trading purposes Greenland is monopolized by the Danish Government; its Esquimaux and mixed population amount to about 7000 souls. About 1000 Danes reside constantly there for the purpose of conducting the trade, which consists almost exclusively in the exchange of European goods for oil and the skins of seals, reindeer, and a few other animals.

The Esquimaux are not subject to Danish laws, but although proud of their nominal independence they are sincerely attached to the Danes, and with abundant reason; a Lutheran clergyman, a doctor, and a schoolmaster, whose duty it is to give gratuitous instruction and relief, are paid by the Government, and attached to each district; and when these improvident people are in distress, which not unfrequently happens during the long winters, provisions are issued to them free of cost; spirits are strictly prohibited. All of them have become Christians, and many can read and write.

Have we English done more, or as much, for the aborigines in any of our numerous colonies, and especially for the Esquimaux within our own territories of Labrador and Hudson's Bay?

Greenland is divided into two inspectorates, the northern and southern; the inspector of the latter division, Dr. Rink, had arrived at Frederickshaab upon his summer round of visits only the day previous to ourselves. He came on board to call upon me, and after Divine Service I landed, and enjoyed a ramble with him over the moss-clad hills. Our first meeting was in North Greenland in 1848; we had not seen one another since, so we had much to talk about. Dr. Rink is a gentleman of acknowledged talent, a distinguished traveller, and is thoroughly conversant with the sciences of geology and botany.

Unfortunately for me his excellent work on Greenland has not been translated into English. We were kindly permitted to purchase eight tons of coals, and such small things as were required; the only fresh supplies to be obtained besides codfish, which was abundant, consisted of a very few ptarmigan and hares, and a couple of kids; these last are scarce. Some goats exist, but for eight months out of the year they are shut up in a house, and even now—in midsummer—are only let out in the daytime. We also purchased of the Esquimaux some specimens of Esquimaux workmanship, such as models of the native dresses, kayaks, &c., also birds' skins and eggs. I saw fine specimens of a white swan, and of a bird said to be extremely rare in Greenland,—it was a species of grebe, *Podiceps cristatus*, I imagine. Frederickshaab is just now well supplied with wood: besides an unseaworthy brig, the wreck of a large timber-ship lay on the beach, and an abandoned timber-vessel, which was met with between Iceland and Greenland in July by Prince Napoleon, drifted upon the coast 30 miles to the northward in the following September.

CHAPTER II.

Fiskernaes and Esquimaux — The 'Fox' reaches Disco — Disco Fiord — Summer scenery — Waigat Strait — Coaling from the mine — Purchasing Esquimaux dogs — Heavy gale off Upernivik — Melville Bay — The middle ice — The great glacier of Greenland — Reindeer cross the glacier.

23rd July.—SAILED the day before yesterday for Godhaab. The fog was thick, and wind strong and contrary, but the current being favourable we found ourselves off the small out station of Fiskernaes, when early this morning our fore topmast was carried away; this accident induced me to run in and anchor for the purpose of repairing the damage.

After passing within the outer islets the Moravian settlement of Lichtenfels came in view upon the right hand; it consists of a large sombre-looking wooden house over which is a belfry, a smaller wooden house, and about a dozen native huts roofed with sods, and scarcely distinguishable from the ground they stand on, even at a very short distance. The land immediately behind is a barren rocky steep, now just sufficiently denuded of snow to look desolate in the extreme. A strong tide was setting out of the fiord, as we approached and anchored in the rocky little cove of Fiskernaes: here we were not only sheltered from the wind, but the steep dark rocks within a ship's length on each side of us reflected a strong heat, whilst large mosquitoes lost no time in paying us their annoying visits. This remote spot has been visited by the Arctic voyagers Captain Inglefield, R.N., and Dr. Kane, U.S.N., and still more recently by Prince Napoleon. Dr. Kane's account of his visit is full and very interesting. Cod-fishing was now in full activity, and the few men not so employed had gone up the fiord to hunt reindeer.

The solitary dwelling-house belongs of course to the chief trader, and is a model of cleanliness and order; built of wood, it exhibits all the resources of the painter's art; the exterior is a dull red, the window-frames are white, floors yellow, wooden partitions and low ceilings pale blue. The lady of the house had resided here for about eight years, and appeared to us to be, and acknowledged she was, heartily tired of the solitude. She gave me coffee, and some seeds for cultivation at our winter quarters: these were lettuce, spinach, turnips, carraway, and peas, the latter being the common kind used on board ship;

usually they have only produced leaves on this spot, but once the young peas grew large enough for the table. I expressed a wish to see the interior of an Esquimaux tent. Petersen pulled aside the thin membrane of some animal, which hung across a doorway, and served to exclude the wind, but admitted light, for, although past midnight, the sun was up. Some seven or eight individuals lay within closely packed upon the ground; the heads of old and young, males and females, being just visible above the common covering. Going to bed here only means lying down with your clothes on, upon a reindeer skin, wherever you can find room, and pulling another fur-robe over you.

Fiskernaes appeared to be a sunny little nook, yet all the people we saw there were suffering from colds and coughs, and many deaths had occurred during the spring. The boys brought us handfuls of rough garnets, some of them as large as walnuts, receiving with evident satisfaction biscuits in exchange.

By next morning we were able to put to sea, and early on the day following arrived off the large settlement of Godhaab; it is in the "Gilbert Sound" of Davis, and appears in many old charts as Baal's River. Almost adjoining Godhaab is the Moravian settlement of New

Herrnhut. Here it was that Hans Egede, the missionary father of Greenland, established himself in 1721, and thus re-opened the communication between Europe and Greenland, which had ceased upon the extinction of its early Scandinavian settlers in the 14th century.

A few years after Egede's successful beginning, the Moravian mission still existing under the name of New Herrnhut was established. At present the Moravians support four missions in Greenland; they are not subject to the Danish authorities, but are not permitted in any way to trade.

As we were about to enter the harbour, the Danish vessel—the sole object of our visit—came out, so not a moment was lost in sending on board our invalid and our letter-bag, and in landing our coasting pilot. This man had brought us up from Frederickshaab for the very moderate sum of three pounds; he was an Esquimaux, and, as the brother of poor Hans, Dr. Kane's unhappy dog-driver, was received with favour amongst us, and soon won our esteem by his quiet obliging disposition, as also by his ability in the discharge of his duty; he was so keensighted and so vigilant, it was quite a comfort to have him on board during the foggy weather, for he could recognise on the instant every rock or point, even when dimly looming through the mist. We were not long in discovering that his absence was a loss to us.

When passing out to the north of the Kookornen islands, the wind suddenly failed, and at the same time a swell from to seaward reached us; we therefore had considerable difficulty in towing the ship clear of the rocks; for nearly half an hour our position was most critical.

July 31*st.*—Anchored at Godhaven (or Lievely), in Disco, for a few hours. I presented a letter from the Directors of the Royal Greenland Commerce to the Inspector of North Greenland, Mr. Olrik, authorising him to furnish us with any needful supplies. Our only wants were sledge-dogs and a native to manage them. We soon obtained ten of the former, but were advised to go into Disco Fiord, where many of the Esquimaux were busy in taking and drying salmon-trout, and where some would most probably be obtained.

I was much pleased with Mr. Olrik's kind reception of me, and soon found him to be not only agreeable but well informed; born in Greenland of Danish parents, he is thoroughly conversant with the language and habits of the

Esquimaux, and has devoted much of his leisure time in collecting rare specimens of the animal, vegetable, and mineral productions of the country. I came away enriched by some fossils from the fossil forest of Atanekerdluk, also with specimens of native coal.

It was here I met with the late commanders of the whalers 'Gipsy' and 'Undaunted,' of Peterhead, which had been crushed by the ice in Melville Bay five or six weeks previously; all the other whalers had returned from the north along the pack edge, and passed south of Disco. They said that the ice in Melville Bay was all broken up, and that they thought we should find but little difficulty at this late period in passing through it into the North Water.

Leaving Godhaven in the afternoon with a native pilot, we found ourselves some 10 or 12 miles up Disco Fiord at an early hour next morning. After despatching the pilot to announce our arrival to his countrymen at their fishing station, 7 or 8 miles further up, the Doctor and I landed upon the north side to explore.

The scenery is charming, lofty hills of trap rock, with unusually rich slopes (for the 70th parallel) descending to the fiord, and strewed with boulders of gneiss and granite. We found the blue campanula holding a conspicuous place amongst the wild flowers. I do not know a more enticing spot in Greenland for a week's shooting, fishing, and yachting than Disco Fiord; hares and ptarmigan may be found along the bases of the hills; ducks are most abundant upon the fiord, and delicious salmon-trout very plentiful in the rivers. Formerly Disco was famed for the large size and abundance of its reindeer; but for some unexplained reason they now confine themselves to the mainland.

At this season the natives of Godhaab resort here and enjoy the trout fishery,—it is truly their season of harvest: the weather is pleasant, food delicious and abundant, and the labour an agreeable pastime.

Some kayaks soon came off to the ship, bringing salmon-trout, both fresh and smoked.

A young Esquimaux, named Christian, volunteered his services as our dog-driver, and was accepted; he is about 23 years of age, unmarried, and an orphan. The men soon thoroughly cleansed and cropped him: soap and scissors being novelties to an Esquimaux: they then rigged him in sailor's clothes; he was evidently not at home in them, but was not the less proud of his improved appearance, as reflected in the admiring glances of his countrymen.

We now hastened away to the Waigat Strait to complete our coals. When passing Godhaven, the pilot was launched off our deck in his little kayak without stopping the ship! As a kayak is usually about 18 feet long, 8 inches deep, and only 16 or 17 inches wide, it requires great expertness to perform such a feat without the addition of a capsize.

4th August.—Entered the Waigat yesterday morning, slowly steaming through a sea of glass. Its surface was only rippled by the myriads of eider-ducks which extended over it for several miles: most of them were immature in plumage, and were probably the birds of last year.

After running about 24 miles, towards evening we approached a low range of sandstone cliffs on the Disco shore, in which horizontal seams of coal were seen. Here we anchored, and immediately commenced coaling. It was fortunate we did so, for soon it began to blow hard; and ere noon to-day we were obliged, for the safety of the ship, to leave our exposed anchorage, having however secured eight or nine tons of tolerable coal. Formerly these coal-seams were worked for the supply of the neighbouring settlements, but for several years past it has been found more profitable and convenient to send out coals from Denmark, and thus permit the natives to devote their whole time to the seal-fishery.

The Waigat scenery is unusually grand; the strait varies from 3 to 5 leagues in width; on each side are mountains of 3000 feet in height. The Disco side, upon which we landed, is composed of trap, sandstone appearing only at the beach, and occasionally rising in cliffs to about 100 feet. Upon the moss-clad slopes many fragments of quartz and zeolite were met with. The north end of Disco is almost a precipice to its snow-capped summit, which is 4000 feet high.

5th.—A pleasant fair wind carries us rapidly northward, passing many icebergs. Our rigging is richly garnished with split codfish, which we hoped would dry and keep; but a warm day in Disco Fiord, and much rain with a southerly gale in the Waigat, have destroyed it for our own use. It is however still valuable as food for our dogs. I am very anxious to complete my stock of these our native auxiliaries, as without them we cannot hope to explore all the lands which it is the object of our voyage to search. We could only obtain ten at Godhaven, and require twenty more.

6th.—By Petersen's intimate knowledge of the coast we were enabled to run close in to the little settlement of Proven during the night, and obtain a few dogs and dogs' food. This morning we reached the extreme station of Upernivik, the last trace of civilization we shall meet with for some time. It is in lat. 72¾ N. Here Petersen resided for twelve of the eighteen years he has spent in Greenland, and his unlooked-for reappearance astonished and delighted the small community, more especially Governor Fliescher and his household, who received us with a most hearty welcome.

7th.—Yesterday, when we hove to off Upernivik, the weather was very bad and rapidly growing worse, therefore our stay was limited to a couple of hours. The last letters for home were landed, fourteen dogs and a quantity of seal's flesh for them embarked, and the ship's head was turned seaward.

It was then blowing a southerly gale, with overcast murky sky, and a heavy sea running. When four miles outside the outer island, breakers were suddenly discovered ahead, only just in time to avoid the ledge of sunken rocks upon which the sea was beating most violently. Many such rocks lie at considerable distances beyond the islands which border this coast, and greatly add to the dangers of its navigation. Being now fairly at sea, and the ship under easy sail for the night, I went early to bed in the hope of sleeping. I had been up all the previous night, naturally anxious about the ship threading her way through so many dangers, uncertain about being able to complete the number of our sledge-dogs, and much occupied in closing my correspondence, to which there would be an end for at least a year. All this over, the uncertain future loomed ominously before me. The great responsibilities I had undertaken seemed now and at once to fall with all their weight upon me. A mental whirlpool was the consequence, which, backed by the material storm, and the howling of the wretched dogs in concert on deck, together with the tumbling about of everything below, long kept sleep in abeyance. One thought and feeling predominated: it was gratitude, deep and humble, for the success which had hitherto attended us, and for some narrow escapes which I must ever regard as Providential.

Yesterday's gale has given place to calm foggy weather. An occasional iceberg is seen. The officers amuse themselves in trying new guns, and shooting sea-birds for our dogs.

Governor Fliescher told me yesterday that for the last four weeks southerly winds prevailed, and that only a fortnight ago his boat was unable to reach the Loom Cliffs at Cape Shackleton, 50 miles north of Upernivik, in consequence of the ice being pressed in against the land. I fear these same winds have closed together the ice which occupies the middle of Davis' Strait (hence called the middle ice), so that we shall not be able to penetrate it. However, we are standing out to make the attempt.

To the uninitiated it may be as well to observe that each winter the sea called Baffin's Bay freezes over; in spring this vast body of ice breaks up, and drifting southward in a mass—called the main-pack, or the middle ice—obstructs the passage across from east to west.

The "North Passage" is made by sailing round the north end of this pack; the "Middle Passage," by pushing through it; and the "Southern Passage," by passing round its southern extreme; but seasons do occur when none of these routes are practicable.

It is very remarkable that southward of Disco northerly winds have prevailed. They greatly impeded our progress up Davis' Strait, but we cheered ourselves with the hope that they would effectually clear a path for us across the northern part of Baffin's Bay.

8*th.*—Last night we reached the edge of the middle ice, about 70 miles to the west of Upernivik, and ran southward along its edge all night. This morning, in thick fog, the ship was caught in its margin of loose ice. The fog soon after cleared off, and we saw the clear sea about two miles to the eastward, whilst all to the west was impenetrable closely-packed floe-pieces. After steaming out of our predicament (a matter which we could not accomplish under sail) we ran on to the southward until evening, but found the pack edge still composed of light ice very closely pressed together.

Having now closely examined it for an extent of 40 miles, I was satisfied that we could not force a passage through it across Baffin's Bay, as is frequently done in ordinary seasons: therefore, taking advantage of a fair wind, we steered to the northward, in order to seek an opening in that direction.

12*th.*—We are in Melville Bay; made fast this afternoon to an iceberg, which lies aground in 58 fathoms water, about 2 miles from Browne's Islands, and between them and the great glacier which here takes the place of the coast-line.

We have got thus far without any difficulty, sailing along the edge of the middle ice; but here we find it pressing in against Browne's Islands, and covering the whole bay to the northward, quite in to the steep face of the glacier. This is evidently the result of long-continued southerly winds; but as the ice is very much broken up, we may expect it to move off rapidly before the autumnal northerly winds now due, and these winds invariably remove the previous season's ice. All that we know of Melville Bay navigation in August is derived from the experience of Government and private searching expeditions during eight or nine seasons. My own three previous transits across it were made in this month. The whalers either get through in June or July, or give up the attempt as being too late for their fishing. It frequently happens that they get round the south end of the middle ice, between latitudes 66° and 69° N., and up the west coast of Baffin's Bay late in the season; but we have no accounts of these voyages, nor should I be justified, at this late period of the season, in abandoning the prospect before me, in order to attempt a route which, even if successful, would lengthen our voyage to Barrow's Strait by 700 or 800 miles. We have already passed what is usually the most difficult and dangerous part of the Melville Bay transit.

There is much to excite intense admiration and wonder around us; one cannot at once appreciate the grandeur of this mighty glacier, extending unbroken for 40 or 50 miles. Its sea-cliffs, about 5 or 6 miles from us, appear comparatively low, yet the icebergs detached from it are of the loftiest description. Here, on the spot, it does not seem incorrect to compare the icebergs to mere chippings off its edge, and the floe-ice to the thinnest shavings.

The far-off outline of glacier, seen against the eastern sky, has a faint tinge of yellow: it is almost horizontal, and of unknown distance and elevation.

There is an unusual dearth of birds and seals: everything around us is painfully still, excepting when an occasional iceberg splits off from the parent glacier; then we hear a rumbling crash like distant thunder, and the wave occasioned by the launch reaches us in six or seven minutes, and makes the ship roll lazily for a similar period. I cannot imagine that within the whole compass of nature's varied aspects there is presented to the human eye a scene so well adapted for promoting deep and serious reflection, for lifting the thoughts from trivial things of everyday life to others of the highest import.

The glacier serves to remind one at once of Time and of Eternity—of time, since we see portions of it break off to drift and melt away; and of eternity, since its downward march is so extremely slow, and its augmentations behind so regular, that no change in its appearance is perceptible from age to age. If even the untaught savages of luxuriant tropical regions regard the earth merely as a temporary abode, surely all who gaze upon this ice-overwhelmed region, this wide expanse of "terrestrial wreck," must be similarly assured that here "we have no abiding place."

During daytime the strong glare is very distressing, hence the subdued light of midnight, when the sun just skims along the northern horizon, is much the most agreeable part of the twenty-four hours; the temperature varies between 30° and 40° of Fahrenheit.

The drift-ice of various descriptions about us is constantly in motion under the influence of mysterious surface and under currents (according to their relative depths of floatation), which whirl them about in every possible direction.

To the S.E. are two small islands, almost enveloped in the glacier, and far within it an occasional mountain-peak protrudes from beneath.

From observing closely the variations in the glacier surface, I think we may safely infer that where it lies unbroken and smooth, the supporting land is level; and where much crevassed, the land beneath is uneven. The crevassed parts are of course impassable, but, by following the windings of the smooth surface, I think the interior could be reached. Some attempts to cross the glacier in South Greenland have failed, yet, by studying its character and attending to this remark, I think places might be found where an attempt would succeed. Mr. Petersen tells me that the Esquimaux of Upernivik are unable to account for occasional disappearances and reappearances of immense herds of reindeer, except by assuming that they migrate at intervals to feeding-grounds beyond the glacier, the surface of which he also says is smooth enough in many places even for dog-sledges to travel upon. As there is much uninhabited land both to the northward and southward of Upernivik, I do not see the necessity for this supposition. The habits of the Esquimaux confine them almost exclusively to the islands and seacoasts.

CHAPTER III.

Melville Bay — Beset in Melville Bay — Signs of winter — The coming storm — Drifting in the pack — Canine appetite — Resigned to a winter in the pack — Dinner stolen by sharks — The Arctic shark — White whales and Killers.

15*th Aug.*—THREE days of the most perfect calm have sadly taxed our patience. Lovely bright weather, but scarcely a living creature seen. This afternoon the anxiously-looked-for north wind sprang up, and immediately the light ice began to drift away before it, but it is not strong enough to influence the icebergs, and they greatly retard the clearing-out of the bay. We have noticed a constant wind off the glacier, probably the result of its cooling effect upon the atmosphere; this wind does not extend more than 3 or 4 miles out from it.

16*th.*—One of the loveliest mornings imaginable: the icebergs sparkled in the sun, and the breeze was just sufficiently strong to ripple the patches of dark blue sea; beyond this, there was nothing to cheer one in the prospect from the Crow's-nest at four o'clock; but little change had taken place in the ice; I therefore determined to run back along the pack-edge to the south-westward, in the hope that some favourable change might have taken place further off shore. The barometer was unusually low, yet no indication of any change of weather. A seaman's chest was picked up; it contained only a spoon, a fork, and some tin canisters, and probably drifted here from the southward, where the two whale-ships were crushed in June, affording another proof of the prevalence of southerly winds. As we steamed on, the ice was found to have opened considerably; it fell calm, and mist was observed rolling along the glacier from the southward. By noon a S.E. wind reached us; all sail was set, the leads or lanes of water became wider, and our hopes of speedily crossing Melville Bay rose in proportion as our speed increased. We are pursuing our course without let or hindrance.

17*th*.—The fog overtook us yesterday evening; and at length, unable to see our way, we made fast at eleven o'clock to the ice. The wind had freshened, it was evidently blowing a gale outside the ice. During the night we drifted rapidly together with the ice, and this morning, on the clearing off of the fog, we steamed and sailed on again, threading our way between the floes, which are larger and much covered with *dry* snow. This evening we again made fast, the floes having closed together, cutting off advance and retreat. A wintry night, much wind and snow.

19*th*.—Continued strong S.E. winds, pressing the ice closely together, dark sky and snow; everything wears a wintry and threatening aspect; we are closely hemmed in, and have our rudder and screw unshipped. This recommencement of S.E. winds and rapid ebbing of the small remaining portion of summer makes me more anxious about the future than the present. Yesterday the weather improved, and by working for thirteen hours we got the ship out of her small ice-creek into a larger space of water, and in so doing advanced a mile and a half. It is now calm, but the ice still drifts, as we would wish it, to the N.W. Yesterday we were within 12 miles of the position of the 'Enterprise' upon the same day in 1848, and under very similar conditions of weather and ice also.

20*th*.—No favourable ice-drift: this detention has become most painful. The 'Enterprise' reached the open water upon this day in 1848, within 50 miles of our present position; unfortunately, our prospects are not so cheering. There is no relative motion in the floes of ice, except a gradual closing together, the small spaces and streaks of water being still further diminished. The temperature has fallen, and is usually below the freezing-point. I feel most keenly the difficulty of my position; we cannot afford to lose many more days. Of all the voyages to Barrow Strait, there are but two which were delayed beyond this date, viz., Parry's in 1824, and the 'Prince Albert's' in 1851. Should we not be released, and therefore be compelled to winter in this pack, notwithstanding all our efforts, I shall repeat the trial next year, and in the end, with God's aid, perform my sacred duty.

The men enjoy a game of rounders on the ice each evening; Petersen and Christian are constantly on the look-out for seals, as well as Hobson and Young occasionally; if in good condition and killed instantaneously, the seals float; several have already been shot; the liver fried with bacon is excellent.

Birds have become scarce,—the few we see are returning southward. How anxiously I watch the ice, weather, barometer, and thermometer! Wind from any other quarter than S.E. would oblige the floe-pieces to rearrange themselves, in doing which they would become loose, and then would be our opportunity to proceed.

24*th*.—Fine weather with very light northerly winds. We have drifted 7 miles to the west in the last two days. The ice is now a close pack, so close that one may walk for many miles over it in any direction, by merely turning a little to the right or left to avoid the small water spaces. My frequent visits to the crow's-nest are not inspiriting: how absolutely distressing this imprisonment is to me, no one without similar experience can form any idea. As yet the crew have but little suspicion how blighted our prospects are.

27*th*.—We daily make attempts to push on, and sometimes get a ship's length, but yesterday evening we made a mile and a half! the ice then closed against the ship's sides and lifted her about a foot. We have had a fresh east wind for two days, but no corresponding ice-drift to the west; this is most discouraging, and can only be accounted for by supposing the existence of much ice or grounded icebergs in that direction.

The dreaded reality of wintering in the pack is gradually forcing itself upon my mind,—but I must not write on this subject, it is bad enough to brood over it unceasingly. We can see the land all round Melville Bay, from Cape Walker nearly to Cape York. Petersen is indefatigable at seal-shooting, he is so anxious to secure them for our dogs; he says they must be hit in the head; "if you hit him in the beef that is not good," meaning that a flesh-wound does not prevent their escaping under the ice. Petersen and Christian practise an Esquimaux mode of attracting the seals; they scrape the ice, thus making a noise like that produced by a seal in making a hole with its flippers, and then place one end of a pole in the water and put their mouths close to the other end, making noises in imitation of the snorts and grunts of their intended victims; whether the device is successful or not I do not know, but it looks laughable enough.

Christian came back a few days ago, like a true seal-hunter, carrying his kayak on his head, and dragging a seal behind him. Only two years ago Petersen returned across this bay with Dr. Kane's retreating party; he shot a seal which they devoured raw, and which, under Providence, saved their lives. Petersen is a good ice-pilot, knows all these coasts as well as or better than any man living, and, from long experience and habits of observation, is almost unerring in his prognostications of the weather. Besides his great value to us as interpreter, few men are better adapted for Arctic work,—an ardent sportsman, an agreeable companion, never at a loss for occupation or amusement, and always contented and sanguine. But we have happily many such dispositions in the 'Fox.'

30*th.*—The whole distance across Melville Bay is 170 miles: of this we have performed about 120, 40 of which we have drifted in the last fourteen days. The 'Isabel' sailed freely over this spot on 20th August, 1852; and the 'North Star' was beset on 30th July, 1849, to the southward of Melville Bay, and carried in the ice across it and some 70 or 80 miles beyond, when she was set free on 26th September, and went into winter quarters in Wolstenholme Sound. What a precedent for us!

Yesterday we set to work as usual to warp the ship along, and moved her ten feet: an insignificant hummock then blocked up the narrow passage; as we could not push it before us, a two-pound blasting charge was exploded, and the surface ice was shattered, but such an immense quantity of broken ice came up from beneath, that the difficulty was greatly increased instead of being removed. This is one of the many instances in which our small vessel labours under very great disadvantages in ice-navigation—we have neither sufficient manual power, steam power, nor impetus to force the floes asunder. I am convinced that a steamer of moderate size and power, with a crew of forty or fifty men, would have got through a hundred miles of such ice in less time than we have been beset.

The temperature fell to 25° last night, and the pools are strongly frozen over. I now look matters steadily and calmly in the face; whilst reasonable ground for hope remained I was anxious in the extreme. The dismal prospect of a "winter in the pack" has scarcely begun to dawn upon the crew; however, I do not think they will be much upset by it. They had some exciting foot-races on the ice yesterday evening.

1*st Sept.*—The indications of an approaching S.E. gale are at all times sufficiently apparent here, and fortunately so, as it is the dangerous wind in Melville Bay. It was on the morning of the 30th, before church-time, that they attracted our attention: the wind was very light, but barometer low and falling; very threatening appearances in the S.E. quarter, dark-blue sky, and grey detached clouds slowly rising; when the wind commenced the barometer began to rise. This gale lasted forty-eight hours, and closed up every little space of water; at first all the ice drifted before the wind, but latterly remained stationary. Twenty seals have been shot up to this time.

On comparing Petersen's experience with my own and that of the 'North Star' in 1849, it seems probable that the ice along the shores of Melville Bay, at this season, will drift northward close along the land as far as Cape Parry, where, meeting with a S.W. current out of Whale or Smith's Sound, it will be carried away into the middle of Baffin's Bay, and thence during the winter down Davis' Strait into the Atlantic. From Cape Dudley Digges to Cape Parry, including Wolstenholme Sound, open water remains until October. It is strange that we have ceased to drift lately to the westward.

6*th.*—During the last week we have only drifted 9 miles to the west. Obtained soundings in 88 fathoms; this is a discovery, and not an agreeable one. Of the six or seven icebergs in sight, the nearest are to the west of us; they are very large, and appear to be aground; we approach them slowly. Pleasant weather, but the winds are much too gentle to be of service to us; although the nights are cold, yet during the day our men occasionally do their sewing on deck. Our companions the seals are larger and fatter than formerly, therefore they float when shot; we are disposed to attribute their improved condition to better feeding upon this bank. The dredge brought up some few shell-fish, starfish, stones, and much soft mud.

9*th.*—On this day in 1824 Sir Edward Parry got out of the middle ice, and succeeded in reaching Port Bowen. To continue hoping for release in time to reach Bellot Strait would be absurd; yet to employ the men we continue our preparation of tents, sledges, and gear for travelling. Two days ago the ice became more slack than usual, and a long lane opened; its western termination could not be seen from aloft. Every effort was made to get into this water, and by the aid of steam and blasting-powder we advanced 100 yards out of the intervening 170 yards of ice, when the floes began to close together, a S.E. wind having sprung up. Had we succeeded in reaching the water, I think we should have extricated ourselves completely, and perhaps ere this have reached Barrow Strait, but S.E. and S.W. gales succeeded, and it now blows a S.S.E. gale with sleet.

10*th.*—Young went to the large icebergs to-day; the nearest of them is 250 feet high, and in 83 fathoms water; it is therefore probably aground, except at spring-tide; the floe ice was drifting past it to the westward, and was crushing up against its sides to a height of 50 feet.

13*th*.—Thermometer has fallen to 17° at noon. We have drifted 18 miles to the W. in the last week; therefore our neighbours the icebergs are not always aground, but even when afloat drift more slowly than the light ice. There is a water-sky to the W. and N.W.; it is nearest to us in the direction of Cape York: *could we only advance* 12 *or* 15 *miles in that direction, I am convinced we should be free to steer for Barrow Strait.* Forty-three seals have been secured for the dogs; one dog is missing, the remaining twenty-nine devoured their two days' allowance of seal's flesh (60 or 65 lbs.) in forty-two seconds! it contained no bone, and had been cut up into small pieces, and spread out upon the snow, before they were permitted to rush to dinner; in this way the weak enjoy a fair chance, and there is no time for fighting. We do not allow them on board.

16*th*.—At length we have drifted past the large icebergs, obtaining soundings in 69 fathoms within a mile of them; they must now be aground, and have frequently been so during the last three weeks; and being directly upon our line of drift, are probably the immediate cause of our still remaining in Melville Bay. The ice is slack everywhere, but the temperature having fallen to 3°, new ice rapidly forms, so that the change comes too late. The western limit of the bay—Cape York—is very distinct, and not more than 25 miles from us.

18*th*.—Lanes of water in all directions; but the nearest is half a mile from us. They come too late, as do also the N.W. winds which have now succeeded the fatal south-easters. The temperature fell to 2° below zero last night. We are now at length in the "North Water;" the old ice has spread out in all directions, so that it is only the young ice—formed within the last fortnight—which detains us prisoners here.

The icebergs, the chief cause of our unfortunate detention, and which for more than three weeks were in advance of us to the westward, are now, in the short space of two days, nearly out of sight to the eastward.

The preparations for wintering and sledge-travelling go on with unabated alacrity; the latter will be useful should it become necessary to abandon the ship.

Notwithstanding such a withering blight to my dearest hopes, yet I cannot overlook the many sources of gratification which do exist; we have not only the necessaries, but also a fair portion of the luxuries of ordinary sea-life; our provisions and clothing are abundant and well suited to the climate. Our whole equipment, though upon so small a scale, is perfect in its way. We all enjoy perfect health, and the men are most cheerful, willing, and quiet.

Our "native auxiliaries," consisting of Christian and his twenty-nine dogs, are capable of performing immense service; whilst Mr. Petersen from his great Arctic experience is of much use to me, besides being all that I could wish as an interpreter. Humanly speaking, we were not unreasonable in confidently looking forward to a successful issue of this season's operations, and I greatly fear that poor Lady Franklin's disappointment will consequently be the more severely felt.

We are doomed to pass a long winter of absolute inutility, if not of idleness, in comparative peril and privation: nevertheless the men seem very happy, — thoughtless of course, as true sailors always are.

We have drifted off the bank into much deeper water, and suppose this is the reason that seals have become more scarce.

22*nd.*—Constant N.W. winds continue to drift us slowly southward. Strong indications of water in the N.W., W., and S.E.; its vicinity may account for a rise in the temperature, without apparent cause, to 27° at noon to-day.

The newly formed ice affords us delightful

walking; the old ice on the contrary is covered with a foot of soft snow. We have no shooting; scarcely a living creature has been seen for a week.

24*th*.—Yesterday I thought I saw two of our men walking at a distance, and beyond some unsafe ice, but on inquiry found that all were on board: Petersen and I set off to reconnoitre the strangers; they proved to be bears, but much too wary to let us come within shot. It was dark when we returned on board after a brisk walk over the new ice. The calm air felt agreeably mild. We were without mittens; and but that the breath froze upon mustachios and beard, one could have readily imagined the night was comfortably warm. The thermometer stood at +5°.

To-day when walking in a fresh breeze the wind felt very cold, and kept one on the look-out for frost-bites, although the thermometer was up to 10°. Games upon the ice and skating are our afternoon amusements, but we also have some few lovers of music, who embrace the opportunity for vigorous execution, without fear of being reminded that others may have ears more sensitive and discriminating than their own.

26*th*.—The mountain to the north of Melville Bay, known as the 'Snowy Peak,' was visible yesterday, although 90 miles distant; I have calculated its height to be 6000 feet. A raven was shot to-day.

27*th*.—Our salt meat is usually soaked for some days before being used; for this purpose it is put into a net, and lowered through a hole in the ice; this morning the net had been torn, and only a fragment of it remained! We suppose our twenty-two pounds of salt meat had been devoured by a shark; it would be curious to know how such fare agrees with him, as a full meal of salted provision will kill an Esquimaux dog, which thrives on almost anything. I used to remonstrate upon the skins of sea-birds being given to our dogs, but was told the feathers were good for them! Here all sea-birds are skinned before being cooked, otherwise our ducks, divers, and looms would be uneatably fishy. A well-baited shark-hook has been substituted for the net of salt meat; I much wish to capture one of the monsters, as wonderful stories are told us of their doings in Greenland: whether they are the white shark or the basking shark of natural history I cannot find out. It is only of late years that the shark fishery has been carried on to any extent in Greenland; they are captured for the sake of their livers, which yield a considerable quantity of oil. It has very recently been ascertained that a valuable substance resembling spermaceti may be expressed from the carcase, and for this purpose powerful screw presses are now employed. In early winter the sharks are caught with hook and line through holes in the ice.

The Esquimaux assert that they are insensible to pain; and Petersen assures me he has plunged a long knife several times into the head of one whilst it continued to feed upon a white whale entangled in his net!! It is not sufficient to drive them away with sundry thrusts of spears or knives, but they must be towed away to some distance from the nets, otherwise they will return to feed. It must be remembered that the brain of a shark is extremely small in proportion to the size of its huge head. I have seen bullets fired through them with very little apparent effect; but if these creatures *can* feel, the devices practised upon them by the Esquimaux must be cruel indeed.

It is only in certain localities that sharks are found, and in these places they are often attracted to the nets by the animals entangled in them. The dogs are not suffered to eat either the skin or the head, the former in consequence of its extreme roughness, and the latter because it causes giddiness and makes them sick.

The nets alluded to are set for the white whale or the seal; if for the former, they are attached to the shore and extend off at right angles so as to intercept them in their autumnal southern migration, when they swim close along the rocks to avoid their direst foe, the grampus, or killer, of sailors, the *Delphinus orca* of naturalists. When the white whale is stopped by the net it often appears at first to be unconscious of the fact, and continues to swim against it, affording time for the approach of the boat and deadly harpoon from behind. If entangled in the net a very short time suffices to drown them, as, like all the whale tribe, they are obliged to come to the surface to breathe.

The killer is also a cetacean of considerable size, 15 to 20 feet in length, but of very different habits; it is very swift, is armed with powerful teeth, and is gregarious. When in sufficient numbers they even attack the whale, impeding his progress by fastening on his fins and tail. In summer they appear in the Greenland seas, and the seals instantly seek refuge from them in the various creeks and inner harbours; and the Esquimaux hunter in his frail kayak, when he sees the huge pointed dorsal fin swiftly cleaving the surface of the sea, is scarcely less anxious to shun such dangerous company. With such stories as these Petersen beguiles the time; I never tire of listening to them, and now amuse myself in jotting scraps of them down.

CHAPTER IV.

Snow crystals — Dog will not eat raven — An Arctic school — The dogs invade us — Bear-hunting by night — Ice-artillery — Arctic palates — Sudden rise of temperature — Harvey's idea of a sortie.

3rd Oct.—SEPTEMBER has passed away and left us as a legacy to the pack; what a month have we had of anxious hopes and fears!

Up to the 17th S.E. winds prevailed, forcing the ice into a compact body, and urging it north-westward; subsequently N.W. winds set in, drifting it southward, and separating the floe-pieces; but the change of wind being accompanied by a considerable fall of temperature, they were either quickly cemented together again or young ice formed over the newly opened lanes of water, almost as rapidly as the surface of the sea became exposed. During the month the thermometer ranged between + 36° and − 2°. Two more bears and a raven have been seen. A wearied ptarmigan alighted near the ship, but before it could take wing again the dogs caught it, and scarcely a feather remained by the time I could rush on deck.

Our beautiful little organ was taken out of

its case to-day, and put up on the lower deck; the men enjoy its pleasing tones, whilst Christian unceasingly turns the handle in a state of intense delight; he regards it with such awe and admiration, and is so entranced, that one cannot help envying him; of course he never saw one before. The instrument was presented by the Prince Consort to the searching vessel bearing his name which was sent out by Lady Franklin in 1851; it is now about to pass its third winter in the frozen regions.

Two dogs ran off yesterday, in the vain hope, I suppose, of bettering their condition,—we only feed them three times a week at present: they returned this morning.

Seals are daily seen upon the new ice, but in this doubtful sort of light they are extremely timid, therefore our sportsmen cannot get within shot. The bears scent or hear our dogs, and so keep aloof; even the shark has deserted us, the bait remains intact. The snow crystals of last night are extremely beautiful; the largest kind is an inch in length; its form exactly resembles the end of a pointed feather. Stellar crystals two-tenths of an inch in diameter have also fallen; these have six points, and are the most exquisite things when seen under a microscope. I remember noticing them at Melville Island in

March, 1853, when the temperature rose to +8°; as these were formed last night between the temperatures of +6° and +12°, it would appear that the form is due to a certain fixed temperature. In the sun, or even in moonlight, all these crystals glisten most brilliantly; and as our masts and rigging are abundantly covered with them, the 'Fox' never was so gorgeously arrayed as she now appears.

13*th*.—One day is very like another; we have to battle stoutly with monotony; and but that each twenty-four hours brings with it necessary though trivial duties, it would be difficult to remember the date. We take our guns and walk long distances, but see nothing. Two of the dogs go hunting on their own account, sometimes remaining absent all night. What they find or do is a mystery. The weather is generally calm and cold,—very favourable for freezing purposes at all events,—for the ice of only three weeks' growth is two feet thick.

I hardly expect any considerable disruption of the ice before the general break-up in the spring, yet we do not trust any of our provisions upon it, nor is it sufficiently still to set up a magnetic observatory, for which purpose the instruments have been supplied to us.

Petersen still hopes we may escape and get into Upernivik, as the sea is not permanently frozen over there before December. I am surprised to hear that eagles have been seen so far north as Upernivik, although it is but twice in twenty-four years that specimens have been noticed there. In Richardson's 'Fauna Boreali Americana' the extreme northern limit of these birds is given as 66°; but Upernivik is in 72¾°.

A few bear and fox tracks have been seen, but no living creatures for several days, except a flock of ducks hastening southward and a solitary raven.

It is said that Esquimaux dogs will eat everything except fox and raven. There are exceptions, however; one of ours, old "Harness Jack," devoured a raven with much gusto some days ago. All the other dogs allowed their harness to be taken off when they were brought on board; but old Jack will not permit himself to be unrobed; when attempted he very plainly threatens to use his teeth. This canine oddity suddenly became immensely popular, by constituting himself protecting head of the establishment when one of his tribe littered; he took up a most uncomfortable position on top of the family cask (our *impromptu* kennel), and prevented the approach of all the other dogs; but for his timely interference on behalf of the poor little puppies, I verily believe they would all have been stolen and devoured! Dogs may do even worse than eat raven.

I have attempted some experiments for the purpose of determining the mean hourly change of oscillation of a pendulum due to the earth's diurnal motion; but as mine was only 11½ feet in length, I failed of any approach to accuracy. The mean of several observations gave 17° 47′, whereas the change due to our latitude is about 14° 30′. A single experiment gave 14° 10′, and this was the longest in point of time of any of them, the pendulum having swung for thirty-six minutes.

24th.—Furious N.W. and S.E. gales have alternated of late; the ship is housed over, to keep out the driving snow; so high is the snow carried in the air that a little box perforated with small holes and triced up 50 feet high is soon filled up; this box is supplied morning and evening with a piece of prepared paper to detect the presence and amount of ozone in the atmosphere; it is a peculiar pet of the Doctor's.

At eight o'clock this evening I noticed the falling of a very brilliant meteor; it passed through the constellation of Cassiopœia in a N.N.E. direction before terminating its visible existence, which it did very much like a huge rocket; the flash was so brilliant that a man whose back was turned to it mistook the illumination for lightning.

26*th*.—Our school opened this evening, under the auspices of Dr. Walker. He reports eight or nine pupils, and is much gratified by their zeal. At present their studies are limited to the three R's—reading, 'riting, and 'rithmetic. They have asked him to read and explain something instructive, so he intends to make them acquainted with the trade-winds and atmosphere. This subject affords an opportunity of explaining the uses of our thermometer, barometer, ozonometer, and electrometer, which they see us take much interest in. It is delightful to find a spirit of inquiry amongst them. Apart from scholastic occupation, I give them healthful exercise in spreading a thick layer of snow over the deck, and encasing the ship all round with a bank of the same material.

28*th*.—Midnight. This evening, to our great astonishment, there occurred a disruption and movement of the ice within 200 yards of the ship. The night was calm; the reflection of a bright moon, aided by the more than ordinary brilliancy of the stars upon the snowy expanse, made it appear to us almost daylight. As I sit now in my cabin I can distinctly hear the ice crushing; it resembles the continued roar of distant surf, and there are many other occasional sounds; some of them remind one of the low moaning of the wind, others are loud and harsh, as if trains of heavy waggons with ungreased axles were slowly labouring along. Upon a less-favoured night these sounds might be appalling; even as it is they are sufficiently ominous to invite reflection. Cape York has been in sight for some days past.

29th.—Another heavenly night, and still greater ice disturbance; some of the crushed-up pieces are nearly four feet thick. The currents, icebergs, and changes of temperature, may contribute to this ice action; but I think the tides are the chief cause, and for these reasons: that it wants but two days to the full moon, and that the ice-movements are almost confined to the night, and change their direction morning and evening. Now we know that the night-tides in Greenland greatly exceed the day-tides. One thing is evident—the weather continues calm, therefore the winds are not concerned in the matter.

2nd Nov.—Having observed some days ago that a few of the dogs were falling away—from some cause or other not having put on their winter clothing before the recent cold weather set in—they were all allowed on board, and given a good extra meal. Since then we can scarcely keep them out. One calm night they made a charge, and boarded the ship so suddenly that several of the men rushed up, very scantily clothed, to see what was the matter. Vigorous measures were adopted to expel the intruders, and there was desperate chasing round the deck with broomsticks, &c. Many of them retreated into holes and corners, and two hours elapsed before they were all driven out; but though the chase was hot, it was cold enough work for the half-clad men.

Sailors use quaint expressions. The nightly foraging expeditions are called "sorties;" they point out to me the various corners between decks where the "ice corrodes," *i.e.* the moisture condenses and forms frost; a ramble over the ice is called "a bit of a peruse." I presume this indignity is offered to the word perambulation.

There was a very sudden call "to arms" to-night. Whether sleeping, prosing, or schooling, every one flew out upon the ice on the instant, as if the magazine or the boiler was on the point of explosion. The alarm of "A bear close-to, fighting with the dogs," was the cause. The luckless beast had approached within 25 yards of the ship ere the quartermaster's eye detected his indistinct outline against the snow; so silently had he crept up that he was within 10 yards of some of the dogs. A shout started them up, and they at once flew round the bear and embarrassed his retreat. In crossing some very thin ice he broke through, and there I found him surrounded by yelping dogs. Poor fellow! Hobson, Young, and Petersen had each lodged a bullet in him; but these only seemed to increase his rage. He succeeded in getting out of the water, when, fearing harm to the numerous bystanders and dogs, or that he might escape, I fired, and luckily the bullet passed through his brain. He proved to be a full-grown male, 7 feet 3 inches in length. As we all aided in the capture, it was decided that the skin should be offered to Lady Franklin.

The carcase will feed our dogs for nearly a month; they were rewarded on the spot with the offal. All of them, however, had not shown equal pluck; some ran off in evident fright, but others showed no symptom of fear, plunging or falling into the water with Bruin. Poor old Sophy was amongst the latter, and received a deep cut in the shoulder from one of his claws. The authorities have prescribed double allowance of food for her, and say she will soon recover.

For the few moments of its duration the chase and death was exciting. And how strange and novel the scene! A misty moon affording but scanty light—dark figures gliding singly about, not daring to approach each other, for the ice trembled under their feet—the enraged bear, the wolfish howling dogs, and the bright flashes of the deadly rifles.

3rd.—I remained up the greater part of last night taking observations, for the evening mists had passed away, and a lovely moon reigned over a calm enchanting night; through a powerful telescope she resembled a huge frosted-silver melon, the large crater-like depression answering to that part from which the footstalk had been detached. Not a sound to break the stillness around, excepting when some hungry dog would return to the late battlefield to gnaw into the bloodstained ice.

On the 1st the sun paid us his last visit for the year, and now we take all our meals by lamplight.

5th.—In order to vary our monotonous routine, we determined to celebrate the day; extra grog was issued to the crew, and also for the first time a proportion of preserved plum-pudding. Lady

Franklin most thoughtfully and kindly sent it on board for occasional use. It is excellent.

This evening a well-got-up procession sallied forth, marched round the ship with drum, gong, and discord, and then proceeded to burn the effigy of Guy Fawkes. Their blackened faces, extravagant costumes, flaring torches, and savage yells, frightened away all the dogs; nor was it until after the fireworks were set off and the traitor consumed that they crept back again. It was school-night, but the men were up for fun, so gave the Doctor a holiday.

12*th*.—Yesterday I had the good fortune to shoot two seals; they were very fat, and their stomachs were filled with shrimps. To-day Young and Petersen shot three more, and many others have been seen. This is cheering, and entices people out for hours daily. There is just enough movement in the ice to keep a few narrow lanes and small pools of water open; the floes or fields of ice are more inclined to spread out from each other than to close. We have latterly been drifting before northerly winds.

16*th*.—A renewal of ice-crushing within a few hundred yards of us. I can hear it in my bed. The ordinary sound resembles the roar of distant surf breaking heavily and continuously; but when heavy masses come in collision with much impetus, it fully realizes the justness of Dr. Kane's descriptive epithet, "ice-artillery." Fortunately for us, our poor little 'Fox' is well within the margin of a stout old floe: we are therefore undisturbed spectators of ice-conflicts, which would be irresistible to anything of human construction. Immediately about the ship all is still, and, as far as appearances go, she is precisely as she would be in a secure harbour—housed all over, banked up with snow to her gunwales. In fact, her winter plumage is so complete that the masts alone are visible. The deck and the now useless skylights are covered with hard snow. Below hatches we are warm and dry; all are in excellent health and spirits, looking forward to an active campaign next winter. God grant it may be realized!

Yesterday Young shot the fiftieth seal, an event duly celebrated by our drinking *the* bottle of champagne which had been set apart in more hopeful times to be drunk on reaching the North Water—that unhappy failure, the more keenly felt from being so very unexpected.

Petersen saw and fired a shot into a narwhal, which brought the blubber out. When most Arctic creatures are wounded in the water, blubber more frequently than blood appears, particularly if the wound is superficial—it spreads over the surface of the water like oil. Bills of fare vary much, even in Greenland. I have inquired of Petersen, and he tells me that the Greenland Esquimaux (there are many Greenlanders of Danish origin) are not agreed as to which of their animals affords the most delicious food; some of them prefer reindeer venison, others think more favourably of young dog, the flesh of which, he asserts, is "just like the beef of sheep." He says a Danish captain, who had acquired the taste, provided some for his guests, and they praised his *mutton!* after dinner he sent for the skin of the animal, which was no other than a large red dog! This occurred in Greenland, where his Danish guests had resided for many years, far removed from European *mutton.* Baked puppy is a real delicacy all over Polynesia: at the Sandwich Islands I was once invited to a feast, and had to feign disappointment as well as I could when told that puppy was so extremely scarce it could not be procured in time, and therefore sucking-pig was substituted!

19*th.*—A heavy southerly gale has increased the ice movements; happily we are undisturbed. As Young was seated under the lee of a hummock, watching for seals to pop up to breathe, the strong ice under him suddenly cracked and separated! He escaped with a ducking, and was just able to reach his gun from the bank ere it sank through the mixture of snow and water.

Yesterday we were all out; I saw only one seal, but was refreshed by the sight of a dozen narwhals. It is a positive treat to see a living creature of any kind. The only birds which remain are dovekies, but they are scarce, and, being white, are very rarely visible.

The dogs are fed every second day, when 2 lbs. of seal's flesh—previously thawed when possible—is given to each; the weaker ones get additional food, and they all pick up whatever scraps are thrown out; this is enough to sustain, but not to satisfy them, so they are continually on the look-out for anything eatable. Hobson made one very happy without intending it; he meant only to give him a kick, but his slipper, being down at heel, flew off, and away went the lucky dog in triumph with the prize, which of course was no more seen.

Two large icebergs drift in company with us; our relative positions have remained pretty nearly the same for the last month.

23*rd*.—A heavy gale commenced at N.E. on the 21st, and continued for thirty-six hours unabated in force, but changed in direction to S.S.W. It appears to have been a revolving storm, moving to the N.W. Yesterday, as the wind approached S.E., the temperature rose to + 32°; the upper deck sloppy; the lower deck temperature during Divine Service was 75°!! As the wind veered round to the S.S.W., the wind moderated, and temperature fell; this evening it is − 7°. How is it that the S.E. wind has brought us such a very high temperature? Even if it traversed an unfrozen sea it could not have derived from thence a higher temperature than 29°. Has it swept across Greenland — that vast superficies partly enveloped in glacier, partly in snow? No, it must have been borne in the higher regions of the atmosphere from the far south, in order to mitigate the severity of this northern climate.

Petersen tells me the same warm S.E. wind suddenly sweeps over Upernivik in midwinter, bringing with it abundance of rain; and that it always shifts to the S.W., and then the temperature rapidly falls: this is precisely the change we have experienced in lat. 75°. I believe a somewhat similar, but less remarkable, change of temperature was noticed in Smith's Sound, lat. 78¾° N.

25th.—Mild, "Madeira weather," as Hobson calls it, temperature up to + 7°. By my desire

MOONLIGHT IN THE ARCTIC REGIONS.

Drawn by Captain May.

Dr. Walker is occupied in making every possible experiment upon the freezing of salt water; the first crop of ice is salt, the second less so, the third produces drinkable water, and the fourth is fresh. Frosty efflorescence appears upon ice formed at low temperatures in calm weather—it is brine expressed by the act of freezing. We need not wonder that dogs, when driven hard over this ice, which soon cuts their feet, suffer intense pain, and often fall down in fits; nor that snow, falling upon young (sea) ice, wholly or partially thaws, even when the temperature is but little above zero; when near the freezing-point the young ice thus coated over becomes sludgy and unsafe.

29*th*.—Keen, biting, N.W. winds. No cracks in the ice, therefore no seals. Grey dawn at ten o'clock, and dark at two. The moon is everywhere the sailor's friend, she is a source of comfort to us here. Nothing to excite conversation, except an occasional inroad of the dogs in search of food; this generally occurs at night. Whenever the deck-light which burns under the housing happens to go out, they scale the steep snow banking, and rush round the deck like wolves. "Why, bless you, Sir, the wery moment that there light goes out, and the quartermaster turns his back, they makes a regular sort*ee*, and in they all comes." "But *where do* they come in, Harvey?" "Where, Sir? why everywheres; they makes no more to do, but in they comes, clean over all." Not long ago old Harvey was chief quartermaster in a line-of-battle ship, and a regular magnet to all the younger midshipmen. He would spin them yarns by the hour during the night-watches about the wonders of the sea, and of the Arctic regions in particular—its bears, its icebergs, and still more terrific "auroras, roaring and flashing about the ship enough to frighten a fellow"!

30*th*.—Severe cold has arrived with the full moon; eight days ago the thermometer stood at the freezing-point, it is now 64° below it! So dark is it now that I was able to observe an eclipse of Jupiter's first satellite before three o'clock to-day. For the last two months we have drifted freely backwards and forwards before N.W. and S.E. winds; each time we have gained a more off-shore position, being gradually separated further and further from the land by fresh growths of ice, which invariably follow up every ice-movement. In this manner we have been thrust out to the S.W. 80 miles from the nearest land, and into that free space which in autumn was open water, and which we then vainly struggled to reach.

That the ice has been most free to move in this direction is additional evidence of the recent proximity of an open sea, and shows that in all probability—I had almost said certainty—we should have sailed, or at least drifted into it, had it not been for those enemies to all progress, the grounded bergs.

CHAPTER V.

Burial in the pack — Musk oxen in lat. 80° north — Thrift of the Arctic fox — The aurora affects the electrometer — An Arctic Christmas — Sufferings of Dr. Kane's deserters — Ice acted on by wind only — How the sun ought to be welcomed — Constant action of the ice — Return of the seals — Revolving storm.

4th Dec.—I HAVE just returned on board from the performance of the most solemn duty a commander can be called upon to fulfil. A funeral at sea is always peculiarly impressive; but this evening at seven o'clock, as we gathered around the sad remains of poor Scott, reposing under an Union Jack, and read the Burial Service by the light of lanterns, the effect could not fail to awaken very serious emotions.

The greater part of the Church Service was read on board, under shelter of the housing; the body was then placed upon a sledge, and drawn by the messmates of the deceased to a short distance from the ship, where a hole through the ice had been cut: it was then "committed to the deep," and the Service completed. What a scene it was! I shall never forget it. The lonely 'Fox,' almost buried in snow, completely isolated from the habitable

A FUNERAL ON THE ICE—THE EFFECT OF PARASELENA (MOCK MOONS).

Drawn by Captain May

world, her colours half-mast high, and bell mournfully tolling; our little procession slowly marching over the rough surface of the frozen sea, guided by lanterns and direction-posts, amid the dark and dreary depth of Arctic winter; the death-like stillness, the intense cold, and threatening aspect of a murky, overcast sky; and all this heightened by one of those strange lunar phenomena which are but seldom seen even here, a complete halo encircling the moon, through which passed a horizontal band of pale light that encompassed the heavens; above the moon appeared the segments of two other halos, and there were also mock moons or paraselenæ to the number of six. The misty atmosphere lent a very ghastly hue to this singular display, which lasted for rather more than an hour.

Poor Scott fell down a hatchway two days only before his death, which was occasioned by the internal injuries then received; he was a steady serious man; a widow and family will mourn his loss. He was our engine-driver; we cannot replace him, therefore the whole duty of working the engines will devolve upon the engineer, Mr. Brand.

11*th*.—Calm, clear weather, pleasant for exercise, but steadily cold; thermometer varies between – 20° and – 30°. At noon the blush of dawn tints the southern horizon, to the north the sky remains inky blue, whilst overhead it is bright and clear, the stars shining, and the pole-star near the zenith very distinct. Although there is a light north wind, thin mackerel-clouds are passing from south to north, and the temperature has risen 10°.

I have been questioning Petersen about the bones of the musk oxen found in Smith's Sound; he says the decayed skulls of about twenty were found, all of them to the north of the 79th parallel. As they were all without lower jaws, he says they were killed by Esquimaux, who leave upon the spot the skulls of large animals, but the weight of the lower jaw being so trifling it is allowed to remain attached to the flesh and tongue. The skull of a musk ox with its massive horns cannot weigh less than 30 lbs.

Although it has been abundantly proved by the existence of raised beaches and fossils, that the shores of Smith's Sound have been elevated within a comparatively recent geological period, yet Petersen tells me that there exist numerous ruins of Esquimaux buildings, probably one or two centuries old, all of which are situated upon very low points, only just sufficiently raised above the reach of the sea; such sites, in fact, as would at present be selected by the natives. These ruins show that no perceptible change has taken place in the relative level of sea and land since they were originally constructed. At Petersen's Greenland home, Upernivik, the land has sunk, as is plainly shown by similar ruins over which the tides now flow.

Anything which illustrates the habits of animals in such extremely high latitudes I think is most interesting; their instincts must be quickened in proportion as the difficulty of subsisting increases. Foxes, white and blue, are very numerous; all the birds are merely summer visitors, therefore the hare is the only creature remaining upon which foxes can prey; but the hares are comparatively scarce, how then do the foxes live for eight months of each year? Petersen thinks they store up provisions during the summer in various holes and crevices, and thus manage to eke out an existence during the dark winter's season; he once saw a fox carry off eggs in his mouth from an eider-duck's nest, one at a time, until the whole were removed; and in winter he has observed a fox scratch a hole down through very deep snow, to a câche of eggs beneath.

The men are exercised at building snow huts; for winter or early spring travelling, this knowledge is almost indispensable. Upon a calm day the temperature of the external air being − 33°, within a snow hut the thermometer stood 17° higher, this important difference being due to the transmission of heat through the ice from the sea beneath.

Evaporation goes on through ice from the water underneath it. The interior of each snow hut is coated with crystals, and the ice upon which the huts are built is four feet thick, but when no longer in contact with water I cannot discover any evaporation from ice. For instance, a canvas screen on deck which became wet by the sudden thaw last month still remains frozen stiff.

14*th.*—Of late there has been much damp upon the lower deck. This has now been remedied by enclosing the hatchway within a commodious snow-porch, which serves as a condenser for the steam and vapour from the inhabited deck below.

19*th.*—Light N.W. winds, with occasional mists; the temperature is comparatively mild: −12° to −25°.

It is now the time of spring-tides; they cause numerous cracks in the ice, but why so, at such a great distance from the land, I cannot explain. The three nearest points of land are respectively 110, 140, and 180 miles distant from us.

Much aurora during the last two days. Yesterday morning it was visible until eclipsed by the day-dawn at 10 o'clock. Although we could no longer see it, I do not think it ceased; very thin clouds occupied its place, through which, as through the aurora, stars appeared scarcely dimmed in lustre. I do not imagine that aurora is ever visible in a *perfectly* clear atmosphere. I often observe it just silvering or rendering luminous the upper edge of low fog or cloud banks, and with a few vertical rays feebly vibrating.

Last evening Dr. Walker called me to witness his success with the electrometer. The electric current was so very weak that the gold-leaves diverged at regular intervals of four or five seconds. Some hours afterwards it was strong enough to *keep* them diverged.

21*st.*—Mid-winter day. Out of the Arctic regions it is better known as the *shortest* day. At noon we could just read type similar to the leading article of the 'Times.' Few people could read more than two or three lines without their eyes aching.

27*th.*—Our Christmas was a very cheerful merry one. The men were supplied with several additional articles, such as hams, plum-puddings, preserved gooseberries and apples, nuts, sweetmeats, and Burton ale. After Divine Service they decorated the lower deck with flags, and made an immense display of food. The officers came down with me to see their preparations. We were really astonished! Their mess-tables were laid out like the counters in a confectioner's shop, with apple and gooseberry tarts, plum and sponge-cakes in pyramids, besides various other unknown puffs, cakes, and loaves of all sizes and shapes. We bake all our own bread, and excellent it is. In the background were nicely-browned hams, meat-pies, cheeses, and other substantial articles. Rum and water in wine-glasses and plum-cake was handed to us: we wished them a happy Christmas, and complimented them on their taste and spirit in getting up such a display. Our silken sledge-banners had been borrowed for the occasion, and were regarded with deference and peculiar pride.

In the evening the officers were enticed down amongst the men again, and at a late hour I was requested, as a great favour, to come down and see how much they were enjoying themselves. I found them in the highest good humour with themselves and all the world. They were perfectly sober, and singing songs, each in his turn. I expressed great satisfaction at having seen them enjoying themselves so much and so rationally, I could therefore the better describe it to Lady Franklin, who was so deeply interested in everything relating to them. I drank their healths, and hoped our position next year would be more suitable for our purpose. We all joined in drinking the healths of Lady Franklin and Miss Cracroft, and amid the acclamations which followed I returned to my cabin, immensely gratified by such an exhibition of genuine good feeling, such veneration for Lady Franklin, and such loyalty to the cause of the expedition. It was very pleasant also that they had taken the most cheering view of our future prospects. I verily believe I was the happiest individual on board that happy evening.

Our Christmas-box has come in the shape of northerly winds, which bid fair to drift us southward towards those latitudes wherein we hope for liberation next spring from this icy bondage.

28*th*.—We have been in expectation of a gale all day. This evening there is still a doubtful sort of truce amongst the elements. Barometer down to 28·83; thermometer up to +5°, although the wind has been strong and steady from the

N. for twenty-four hours, low scud flying from the E., snow constantly falling. An hour ago the wind suddenly changed to S.S.E.; the snowing has ceased; thermometer falls and barometer rises.

2nd Jan. 1858.—New Year's day was a second edition of Christmas, and quite as pleasantly spent. We dwelt much upon the anticipations of the future, being a more agreeable theme than the failure of the past. I confess to a hearty welcome for the new year—anxious, of course, that we may escape uninjured, and sufficiently early to pursue the object of our voyage.

Exactly at midnight on the 31st December the arrival of the new year was announced to me by our band—two flutes and an accordion—striking up at my door. There was also a procession, or perhaps I should say a continuation of the band; these performers were grotesquely attired, and armed with frying-pans, gridirons, kettles, pots, and pans, with which to join in and add to the effect of the *other* music!

We have a very level hard walk alongside the ship; it is narrowed to two or three yards in width by a snow-bank four feet high. In the face of this bank some twenty-five holes have been excavated for the dogs, and in them they spend most of their time. It looks very formidable in the moonlight, being a good imitation of a casemated battery.

After our rubber of whist on New Year's night Petersen related to us some of his dreadful sufferings when with the party of deserters from Dr. Kane. They spent the months of October and November in Booth Sound, lat. 77°; all that time upon the verge of starvation, unable to advance or retreat. For these two months they had no other fuel than their small cedar boat, the smoke of which was not endurable in their wretched hut, and without light, for the sun left them in October, unless we except one inch and a half of taper daily, which they made out of a lump of bees'-wax that accidentally found its way into their boat before leaving the ship. In December they regained their vessel. I am surprised that no account of the extreme hardships of this party —so far exceeding that of their shipmates on board—has ever appeared; and I regret it, as I believe they owed their lives to the experience and fidelity of their interpreter Petersen. At first the Esquimaux assisted them; latterly they were quite unable to do so, and became anxious to get rid of their visitors. Observing how weakened they had become, the Esquimaux endeavoured to separate them from their guns and from each other, and even used threatening language.

During December we drifted 67 miles, directly down Baffin's Bay towards the Atlantic, and are now in lat. 74°. Although it is quite impossible to discriminate between the several influences which probably govern our movements, or to ascertain how much is due to each of them—such as the relative positions of ice, land, and open water, winds, currents, and earth's rotation—yet it appears in the present instance that the wind is almost the sole agent in hastening this vast *continent* of ice towards the latitudes of its dissolution. We move before the wind in proportion to its strength: we remain stationary in calm weather. Neither surface nor submarine current has been detected; the large icebergs obey the same influences as the surface ice. We have noticed a slight set to the westward—it is not likely to be produced by current, and may be the result of the earth's motion from west to east.

6th.—Many lanes of water. A seal has been seen, the only one for six weeks. Of the old ice which so closely hemmed us in up to the middle of September, there is hardly any within several miles of us except the large floe-piece we are frozen to. Every crack or lane which opens is quickly covered with young ice, so that it cannot close again; and in this manner the old ice has been spread out. I rejoice in its dispersion!

To-day I put a tumblerful of our strong ale (Allsopp's) on deck to freeze: this was soon effected, the temperature being $-35°$. After bringing it below, and when its temperature had risen to 17°, it was almost all thawed—at 22° it was completely so: it looked muddy, but settled after standing for a couple of hours, when I drank it off, in every way satisfied with my experiment and my beer: it seemed none the worse for its freezing, but rather flat from its long exposure in a tumbler.

17*th*.—Northerly winds blow almost constantly. We have drifted 60 miles since the 1st, and are only 115 miles from Upernivik, —once more upon confines of the habitable world! good light for three hours daily; all this is cheering. We continue our snow-hut practice, and can build one in three-quarters of an hour.

28*th*.—The upper edge of the sun appeared above the horizon to-day, after an absence of eighty-nine days; it was a gladdening sight. I sent for the ship's steward and asked what was the custom on such occasions? "To hoist the colours and serve out an extra half-gill, sir," was the ready reply: accordingly, the Harwich lion soon fluttered in a breeze cool enough to stiffen the limbs of ordinary lions, and in the evening the grog was issued.

30th.—Our messmate Pussy is unwell, and won't eat; in vain has Hobson tempted her with raw seal's flesh, preserved salmon, preserved milk, &c.; at length castor-oil was forcibly administered. Puss is a great favourite. Our finest dog, Sultan, is also sick, and his coat is in bad order; blubber has been prescribed for him;—and poor old Mary has fits, not uncommon after the long winter. Petersen immediately ordered her to be bled by slitting her ear; but Christian, in his fright and haste, cropped the tip of it off. These are our only medical cases. A dovekie, in its white winter plumage, and two seals have been seen lately.

15th Feb.—The returning daylight cheers us up wonderfully—not that we were suffering, either mentally or bodily, but the change is most agreeable; we can take much longer walks than was possible during the dark period. The men have been supplied with muskets, and go out sporting as ardently as schoolboys. I took a long walk towards one of our iceberg companions, but could not quite reach it as weak ice intervened, each step producing an undulation. Finding the point of my knife went through it with but very slight resistance, I gave up the attempt and turned back. The ship's masts were scarcely visible in the distance; almost the whole of the intervening ice was of this winter's growth, and in many places much crushed up.

Daylight reveals to us evidences of vast ice movements having taken place during the dark months when we fancied all was still and quiet; and we now see how greatly we have been favoured, what innumerable chances of destruction we have unconsciously escaped! A few days ago the ice suddenly cracked within ten yards of the ship, and gave her such a smart shock that every one rushed on deck with astonishing alacrity. One of these sudden disruptions occurred between me and the ship when I was returning from the iceberg; the sun was just setting as I found myself cut off. Had I been upon the other side I would have loitered to enjoy a refreshing gaze upon this dark streak of water; but after a smart run of about a mile along its edge, and finding no place to cross, visions of a patrol on the floe for the long night of fifteen hours began to obtrude themselves!

At length I reached a place where the jagged edges of the floes met, so crossed and got safely on board. Nothing was seen during this walk of nearly 25 miles except one seal. Recent gales have drifted us rapidly southward; cracks and lanes are very numerous.

On the 1st a blue (or sooty) fox was shot. Although 130 geographical miles from the nearest land he was very fat, hence we argue dovekies were much more numerous during winter than we supposed. We have often noticed the tracks of foxes following up those of the bears, probably for discarded scraps of the seals upon which they prey. Hobson's favourite dog "Chummie" has returned, after an absence of six days, decidedly hungry, but he can hardly have been without food all that time; some fox may have lured him off. He evinced great delight at getting back, devoted his first attentions to a hearty meal, then rubbed himself up against his own particular associates, after which he sought out and attacked the weakest of his enemies, and, soothed by their howlings, coiled himself up for a long sleep.

1*st March.*—February has been a remarkably mild, cloudy, windy month: the winter temperature may be said to have passed away by the 10th, the average temperature for the first ten days being $-25°$, whilst for the remainder of the month it was $-11°$. Had one fallen asleep for a month at least, he could not reasonably have expected to find a greater change on awaking. Our drift has been also great,—166 miles. We are south of the 70th parallel, and may soon be expelled from our icy home.

On the 24th there was a fearful gale of wind. Had not our housing been very well secured, it must have been blown away. We are preparing for sea, removing the snow from off the deck and round the ship; our skylights have been dug out (in winter they are always covered with a thick layer of snow), and the flood of light which beams down through them is quite charming. How intolerably sooty and smoke-dried everything looks!

On the 27th the first seal of this year was shot; it came in good time, for the fifty-one seals shot in autumn were finished only two days before: our English supply of dogs' food therefore remains almost untouched. Snow was observed to melt against the ship's side exposed to the sun, the thermometer in the shade standing at $-22°$! A very fine dog has died from eating a quantity of salt fish, which he managed to get at although it was supposed to be quite out of his reach.

One of the two large icebergs which commenced this voyage with us last October, in 75½° N., has drifted out of sight to the S.E.; the other one is far off in the N.W. I attribute these increased distances solely to the spreading abroad of the intervening ice.

When we were far north, and probably drifting more slowly than the ice in the stream of Lancaster Sound to the westward of us, the ship's head turned very gradually from right to left, from N.N.W. to W.; when about the parallel of 72° N., we supposed ourselves to be drifting faster than the western ice; in this, as in the previous case, comparing our drift with that of Lieutenant De Haven, the ship's head slowly shifted back to the right as far as W.N.W.; latterly it has not changed at all: we are in a narrower part of Davis' Strait, where the winds probably blow with equal force from shore to shore, and drift the whole pack at an uniform rate.

5*th.*—On the 2nd four fat seals and some dovekies were shot; the largest seal weighed 170 lbs., the smallest 150 lbs.; they were males of the species Phoca hespida, or Phoca fœtida, the latter epithet being by far the most appropriate at this season; the disagreeable odour resembles garlic, and taints the whole animal so strongly that even Esquimaux are nearly overpowered by it: this is almost the only description of seal we have obtained, but the females are at all seasons free from fetor. Several long lanes of water extend at right angles to the straits.

The Doctor has taken a photograph of the ship by the albumen process on glass; the temperature at the time was below zero. Upon the 3rd and 4th a well-marked revolving storm passed nearly over us to the W.N.W.; its extreme diameter was 30 hours, that of the strength of the gale 18 hours; its centre probably passed about one-tenth of its diameter to the S.W. The barometer was rather high, having risen just before the wind commenced at N.E.; but it now fell half an inch in ten hours, and continued to fall until the wind shifted — almost suddenly — through S.E. to S.S.W.; immediately the barometer got up rapidly. As the barometer fell, the temperature rose from zero to +18°, and fell again after the change of wind. This violent storm brought with it a smart hail-shower.

The depression of the ice about the bows, in consequence of a vast accumulation of snowdrift upon it, brought the ship down by the head considerably; to-day this ice suddenly detached itself, and the fore part of the vessel sprang up; she still remains frozen and held down abaft. The snow-banking looks very woe-begone after this *ice-quake;* it inclines out from the ship, and in many places has been prostrated by the shock.

Early on the morning of the 7th the high land of Disco was seen; its distance was upwards of 90 miles.

CHAPTER VI.

A bear-fight — An ice-nip — Strong gales, rapid drift — The 'Fox' breaks out of the pack — Hanging on to floe-edge — The Arctic bear — An ice tournament — The 'Fox' in peril — A storm in the pack — Escape from the pack.

9th March.—A BEAR was seen this morning; but as he was going away from us, the dogs were brought out in the hope that they might keep him at bay until the sportsmen came up. It was very pretty to see them take up the scent, the moment they caught sight of him they set off at full speed. Bruin had seen them first, and increased his pace to a clumsy gallop, yet the dogs were soon around him; he seemed to care but little about them, steadily making off and following the trending of a recently frozen crack in search of clear water, evidently aware that his persecutors would not follow him there.

After five hours all returned on board again; out of the ten dogs four were wounded by his claws,—skin deep only,—but one of the wounds was seven inches in length, as if made with a sharp knife! this was sewed up, the others were merely trimmed, and nature I am informed will do all the rest. It is really wonderful what cures nature and instinct effect: notwithstanding the extreme cold, no external dressings are applied, because the animal must not be prevented from licking its wound. Petersen says this bear must be very thin, else he could not run so fast. I think it very probable that he has been hunted before, and that fear lent him wings. A black whale has been seen.

11*th*.—Two small seals free from taint were shot yesterday, so we had fried liver and steaks for breakfast this morning; both were good, but the steaks were preferred; they were very dark and very tender, had been cut thin, deprived of all fat, and washed in two or three waters to get rid of the blubber.

16*th*.—Several long lanes of water have again opened, but now all of them extend parallel to the direction of the straits; one lane passed within 120 yards of the ship; its extremes are not visible even from aloft; the ice upon its east side has a more rapid southerly motion than that upon its west side.

18*th*.—Last night the ice closed, shutting up our lane, but its opposite sides continued for several hours to move past each other, rubbing off all projections, crushing, and forcing out of water masses four feet thick: although 120 yards distant, this pressure shook the ship and cracked the intervening ice.

I went out with a lantern to see the nip,—it certainly was awe-inspiring; no one in his senses could avoid reflecting upon the inevitable fate of a ship if exposed to such fearful pressure. It is now spring tides.

19*th.*—All yesterday the lane remained open, in the evening it closed with but slight pressure; yet as the opposing fields of ice continued to move in opposite directions, all jagged points were brushed off, and the débris thus formed between their edges presented a heaving surface of ice-masses,—an ice river. On the separation of the floes, mass after mass forced itself up to the surface, until at length all the submerged ice had risen, except such as had been forced quite under their edges. One seldom meets with a cleanly fractured floe-edge, they are usually fringed with crushed-up ice or newly formed sludge.

23*rd.*—Seals and dovekies are now common; the latter have already made considerable advances towards their summer plumage.

Yesterday there was a very heavy S.E. gale; it blew so furiously, and the snow-drift was so dense, that we could neither hear nor see what was going on twenty yards off; at night the ship, becoming suddenly detached from the ice, heeled over to the storm; until the cause was ascertained we thought the ice had broken up, and pressed against the ship. It was not so; but when the weather moderated we found that there had been heavy pressure upon the edge of the floes,—so much, indeed, that the lane of water was now within 70 yards of the 'Fox;' and that ice 4½ feet thick had been crushed during the storm for a distance of about 50 yards.

25th.—Strong N.W. winds lately, the ship rocking to the breeze, and rubbing her poor sides against the ice, producing a creaking sound which is far from pleasant. More ice-squeezing, and a further inroad upon our barrier; it has yielded slightly, nipping the ship, inclining her to port, and lifting her stern about a foot. Occasional groanings within, and surgings of the ice without.

Our boats, provisions, sledges, knapsacks, and equipment, are ready for a hasty departure, —beyond this we can do nothing; as long as our friendly barrier lasts we need not fear, but who can tell the moment it may be demolished, and the ship exposed to destruction? I am scribbling within a foot of the sternpost—in fact there is a notch in my table to receive it; and

I sympathise with its constant groanings; the ice allows it no rest.

27th.—Strong N.W. gale with a return of cold weather. We have drifted 39 miles in the last forty-eight hours! The lane is open; the whole pack appears to have plenty of room to drift, and, I am happy to add, is taking advantage of it,—so much so that the smaller pieces floating freely in the lane can hardly go at the same pace. Our remaining winter companion, the iceberg, was in sight a few days ago, far away to the N.W.; it may be still visible from aloft, but these March gales cut so keenly, that the crow's-nest is but seldom visited.

31st.—Another N.W. gale; it is also spring tides, and this conjunction makes one fearful of ice movement and pressure; but it seems as if the pack had more room to move in, as it does not close much. Seals are often shot, bear tracks are common, and narwhals are frequently seen migrating northward. The bears must prefer the night-time for wandering about, else we could not help seeing them; we often find their tracks within a few hundred yards of the ship.

Although the last, yet this is the coldest day of the month—thermometer down to $-27°$. The mean temperature for March has been unusually high, $-3°$; whilst Lieutenant De Haven's was

– 17°. Notwithstanding that heavy S.E. gales have three times driven us backward, yet we have advanced 100 miles further down Davis' Straits.

6th April.—To-day we enjoy fine weather, the more so since it comes after a tremendous northerly gale of forty-eight hours' duration. Two days ago the friendly old floe, so long our bulwark of defence, was cracked; the lane of water thus formed soon widened to 60 yards, passed within 30 yards of the 'Fox,' and cut off three of our boats. Yesterday morning another crack detached the remaining 30 yards from us, and as it widened the ship swung across the opening; as quickly as we could effect it the ship was again placed alongside the ice and within a projecting point: had it closed only a few feet whilst she lay across the lane, the consequences must have been very serious. Even to effect this slight change of position we were fully occupied for four hours; for the gale blew furiously, and thermometer stood at 12° below zero, and the cold was very much felt; our hawsers were frozen so stiff as to be quite unmanageable, and we were obliged to use the chain cables to warp the ship into safety.

Throughout yesterday the wind continued extremely strong and keen,—fortunately the ice remained perfectly still: our funnels refused to draw up the smoke; so that between the suffocation, the cold, and anxiety lest the ice should move, our Easter Monday was sufficiently miserable. The half of our poor dogs were cut off from the ship by the lane, and continued to howl dismally until late, when the new ice over the lane was strong enough to bear them, and they came across to us.

To-day we have recovered the boats, shot four seals, seen two whales, and much water to the eastward; we are in latitude 67° 18′ N., and highly delighted with the rapidity of our southern drift.

10*th.*—Yesterday evening the setting sun rendered visible the western land, probably Cape Dyer. We have drifted 70 miles in the last week, and are only 18 miles from De Haven's position of escape; but as we are two months earlier, we must expect to be carried farther south.

12*th.*—This morning we drifted ingloriously out of the Arctic regions, and with what very different feelings from those with which we crossed the Arctic circle eight months ago! However, we have not done with it yet; directly the ice lets us go, we will (D. V.) re-enter the frigid zone, and "try again," with, I trust, better success.

A gull and a few terns appeared to-day; these are the first of our summer visitors. The temperature improves; yesterday at one o'clock it was +19° in the shade, +15° in the crow's-nest 70 feet high, and +51° against a black surface exposed to the sun.

16*th*.—Last night a bear came to the ship, was wounded, but escaped; to-day the tracks were followed up for three miles, the bear found, and again wounded—finally the unlucky beast was shot in the water seven miles from the ship; it was lost in consequence of the rapid drifting of the ice, which ran over the floating carcase.

To-night a dense fog-bank rests upon the water to the southward; its upper edge is illuminated by aurora, showing a faint tremulous light.

17*th*.—Another northerly gale; holding fast to the ice with three hawsers; snow-drift limits the view to a couple of miles, so all to the eastward appears water, and to the westward ice.

Last night the ice opened considerably; to secure the ship occupied us for six hours; several of the dogs were again cut off; as the ice they were on was rapidly drifting away, I sent a boat to recover them; it was a difficult and hazardous business, but at length the boat and dogs returned in safety, to my great relief, for it was both dark and late.

18*th*.—Yesterday morning, when I wrote up my journal, I was hoping to hold on quietly to the floe-edge until the wind moderated, when with clear weather we could take advantage of the openings and make some progress towards the clear sea. We were unable to hold on, for the floe-edge broke away, setting us adrift; some time was occupied in fetching off the boats and dogs,—five of the latter unfortunately would not allow themselves to be caught. As speedily as possible the rudder was shipped and sail set, and before three o'clock the ship was running fast to the eastward! During the night the ice closed, and at daylight scarcely any water was visible; with the exception of a couple of icebergs, all the ice in sight was not more than two days old; it mainly owes its origin and rapid growth to the immense quantities of snow blown off the pack.

It still blows hard, and thermometer stands at 11°. A sudden opening of the ice this forenoon allowed us to run a few miles southward, and then it closed again: we are now surrounded by young ice.

20*th*.—We have been carried rapidly past the position where the Arctic discovery ship 'Resolute' was picked up.

Yesterday three bears, a fulmar petrel, and a

snow bunting were seen; to-day a fine bear came within 150 yards, and was shot by our sportsmen; as they were standing round it afterwards upon the ice, a small seal, the only one seen for several days, popped up its head as if to exult over its fallen enemy—it was of course instantly shot: we have learnt to esteem seal's liver for breakfast very highly.

It seems hardly right to call polar bears *land* animals; they abound here,—110 geographical miles from the nearest land,—upon very loose broken-up ice, which is steadily drifting into the Atlantic at the rate of 12 or 14 miles daily; to remain upon it would insure their destruction were they not nearly amphibious; they hunt by scent, and are constantly running across and against the wind, which prevails from the northward, so that the same instinct which directs their search for prey, also serves the important purpose of guiding them in the direction of the land and more solid ice.

I remarked that the upper part of both Bruin's fore-paws were rubbed quite bare: Petersen explains that to surprise the seal a bear crouches down with his fore-paws doubled underneath, and pushes himself noiselessly forward with his hinder legs until within a few yards, when he springs upon the unsuspecting victim, whether

THE GREENLANDER'S SUPPER APPROPRIATED BY A BEAR.

Drawn by Captain May

in the water or upon the ice. The Greenlanders are fond of bear's flesh, but never eat either the heart or liver, and say that these parts cause sickness. No instance is known of Greenland bears attacking men, except when wounded or provoked; they never disturb the Esquimaux graves, although they seldom fail to rob a câche of seal's flesh, which is a similar construction of loose stones above ground.

A native of Upernivik, one dark winter's day, was out visiting his seal-nets. He found a seal entangled, and, whilst kneeling down over it upon the ice to get it clear, he received a slap on the back—from his companion as he supposed; but a second and heavier blow made him look smartly round. He was horror-stricken to see a peculiarly grim old bear instead of his comrade! Without deigning further notice of the man, Bruin tore the seal out of the net and commenced his supper. He was not interrupted; nor did the man wait to see the meal finished.

I had long ago resolved, if we escaped before the 15th, or the 20th April at the latest, to go to Newfoundland to refresh the crew and to refit, even if no damage from the ice should be sustained. In order to do so it would have been necessary for us to visit a Greenland port for a supply of water. We could not have calculated upon much assistance from our engines upon such a voyage, Mr. Brand alone being capable of working the engines, so that ten or twelve hours daily is all the steaming that could have been expected.

But we are still ice-locked, so I purpose going to Holsteinborg in preference to a more southern port, as there we may expect to get reindeer and a small supply of stores suitable to our wants. The whalers sometimes reach Disco in March, Upernivik in May, and the North Water early in June. Unless we should be at once set free, we would not have time to spare for a Newfoundland voyage.

24*th.*—Another anxious week has passed. Latterly we have experienced south-westerly currents similar to those which Parry describes when beset here in June, 1819. To-day we have had a strong S.E. breeze, with snow and dark weather. The wind had greatly moderated when the swell reached us about eight o'clock this evening. It is now ten o'clock; the long ocean swell already lifts its crest five feet above the hollow of the sea, causing its thick covering of icy fragments to dash against each other and against us with unpleasant violence. It is however very beautiful to look upon, the dear old familiar ocean-swell! it has long been a stranger to us, and is welcome in our solitude. If the 'Fox' was as solid as her neighbours, I am quite sure she would enter into this ice-tournament with all their apparent heartiness, instead of audibly making known her sufferings to us. Every considerable surface of ice has been broken into many smaller ones; with feelings of exultation I watched the process from aloft. A floe-piece near us, of 100 yards in diameter, was speedily cracked so as to resemble a sort of labyrinth, or, still more, a field-spider's web. In the course of half an hour the family resemblance was totally lost; they had so battered each other, and struggled out of their original regularity. The rolling sea can no longer be checked; "the pack has taken upon itself the functions of an ocean," as Dr. Kane graphically expresses it.

26*th.*—At sea! How am I to describe the events of the last two days? It has pleased God to accord to us a deliverance in which His merciful protection contrasts—how strongly!—with our own utter helplessness; as if the successive mercies vouchsafed to us during our long long winter and mysterious ice-drift had been concentrated and repeated in a single act. Thus forcibly does His great goodness come home to the mind!

I am in no humour for writing, being still tired, seedy, and perhaps a little sea-sick; at least I have a headache, caused by the rolling of the ship and rattling noise of everything.

On Saturday night, the 24th, I went on deck to spend the greater part of it in watching, and to determine what to do. The swell greatly increased; it had evidently been approaching for hours before it reached us, since it rose in proportion as the ice was broken up into smaller pieces. In a short time but few of them were equal in size to the ship's deck; most of them not half so large. I knew that near the pack-edge the sea would be very heavy and dangerous; but the wind was now fair, and, having auxiliary steam-power, I resolved to push out of the ice if possible.

Shortly after midnight the ship was under sail, slowly boring her way to the eastward; at two o'clock on Sunday morning commenced steaming, the wind having failed. By eight o'clock we had advanced considerably to the eastward, and the swell had become dangerously high, the waves rising ten feet above the trough of the sea. The shocks of the ice against the ship were alarmingly heavy; it became necessary to steer exactly head-on to swell. We slowly passed a small iceberg 60 or 70 feet high; the swell forced it crashing through the pack, leaving a small water-space in its wake, but sufficient to allow the seas to break against its cliffs, and throw the spray in heavy showers quite over its summit.

The day wore on without change, except that the snow and mists cleared off. Gradually the swell increased, and rolled along more swiftly, becoming in fact a very heavy regular sea, rather than a swell. The ice often lay so closely packed that we could hardly force ahead, although the fair wind had again freshened up. Much heavy hummocky ice and large berg-pieces lay dispersed through the pack; a single thump from any of them would have been instant destruction. By five o'clock the ice became more loose, and clear spaces of water could be seen ahead. We went faster, received fewer though still more severe shocks, until at length we had room to steer clear of the heaviest pieces; and at eight o'clock we emerged from the villanous "pack," and were running fast through straggling pieces into a clear sea. The engines were stopped, and Mr. Brand permitted to rest after eighteen hours' duty, for we now have no one else capable of driving the engines.

Throughout the day I trembled for the safety of the rudder, and screw; deprived of the one or the other, even for half an hour, I think our fate would have been sealed; to have steered in any other direction than *against* the swell would have exposed, and probably sacrificed both.

Our bow is very strongly fortified, well plated externally with iron, and so very sharp that the ice-masses, repeatedly hurled against the ship by the swell as she rose to meet it, were thus robbed of their destructive force; they struck us obliquely, yet caused the vessel to shake violently, the bells to ring, and almost knocked us off our legs. On many occasions the engines were stopped dead by ice choking the screw; once it was some minutes before it could be got to revolve again. Anxious moments those!

After yesterday's experience I can understand how men's hair have turned grey in a few hours. Had self-reliance been my only support and hope, it is not impossible that I might have illustrated the fact. Under the circumstances I did my best to insure our safety, looked as stoical as possible, and inwardly trusted that God would favour our exertions. What a release ours has been, not only from eight months' imprisonment, but from the perils of that one day! Had our little vessel been destroyed after the ice broke up, there remained no hope for us. But we have been brought safely through, and are all truly grateful, I hope, and believe.

I grieve to think of poor Lady Franklin and our friends at home. Severely as we have felt the failure of our first season's operations, yet the ordeal is now over with us: not so with her and them,—they have still to experience that bitter disappointment.

Our distance within the pack-edge, where we first made sail yesterday, was 22 miles. Before we got clear of the ice the height of the waves was 13½ feet; after passing through the last of it there was no increase, but the sea was more confused; in fact, within the ice all minor disturbances were quelled or merged into one regular fast-following swell. The ship and her machinery behaved most admirably in the struggle; should I ever have to pass through such an ice-covered, heaving ocean again, let me secure a passage in the 'Fox.'

During our 242 days in the packed-ice of Baffin's Bay and Davis' Straits we were drifted 1194 geographical or 1385 statute miles; it is the longest drift I know of, and our winter, as a whole, may be considered as having been mild, but very windy.

We are steering now for Holsteinborg, where I intend to refit and refresh the crew; it is reputed to be the best place for reindeer upon the coast.

CHAPTER VII.

A holiday in Greenland — A lady blue with cold — The loves of Greenlanders — Close shaving — Meet the whalers — Information of whalers — Disco — Danish hospitality — Sail from Disco — Kindness of the whalers — Danish establishments in Greenland.

Wednesday night, April 28th.—SAFELY anchored at Holsteinborg, and moored to the rocks; a charming change, after our position only a few days back. We have been visited by the Danish residents — the chief trader or governor, the priest, and two others: their latest European intelligence is not more recent than our own, but the Danish ship is hourly expected; she usually leaves Copenhagen about the middle of March.

The winter here has been just the reverse of our own experience; it has been severe in point of temperature, but with very little wind; the land lies buried in snow, and as yet there is no thaw; it is too early for the cod-fishery, and not a single reindeer has been killed throughout the winter! Eider-ducks, looms, and dovekies are abundant, as well as hares and ptarmigan.

29th.—A bright and lovely day. Our poor, half-famished dogs have been landed near the carcases of four whales, so they must be supremely happy. I visited the Governor to-day, and found his little wooden house as scrupulously clean and neat as the houses of the Danish residents in Greenland invariably are. The only ornaments about the room were portraits of his unfortunate wife and two children: they embarked at Copenhagen last year to rejoin him, and the ill-fated vessel has never since been heard of. Poor Governor Elberg is in ill health, and talks of returning home—by *home* he means Denmark, the land of his birth, and where once he had a home.

30*th*.—This is a grand Danish holiday; the inhabitants are all dressed in their Sunday clothes—at least, all who have got a change of garments—and there is both morning and evening service in the small wooden church. As the Governor could not be persuaded to unlock the door of the dance-house, our men returned on board early; yesterday evening they were all on shore, and, with the Esquimaux, were squeezed into this one large room: to be squeezed in a crowd of human beings is positive enjoyment after a winter's isolation such as ours has been. Old Harvey constituted himself master of the ceremonies, and with his flute led the orchestra; it consisted of one other flute and a fiddle: he managed to perch himself above all the rest, at one end of the room, and played with such vigour that our bluejackets and the Esquimaux ladies danced away most furiously for hours. These ladies can dance in the least possible space, their costume being particularly well adapted for the purpose, partaking as it does much more of the "Bloomer" than the "crinoline."

Christian looks immensely happy: his countrymen regard him as a man whose fortune is made, and the women gaze with admiration upon his neat sailor's dress, and his goodnatured full, round face, and huge fat, shining cheeks; Mr. Petersen is in great request to interpret between the English, Danes, and Esquimaux.

7th May.—I intended sailing for Disco this morning, but wind and weather were adverse. We have obtained but little here except water, a tolerable supply of rock cod, some ptarmigan, hares, wildfowl, and a few items of stores. The Governor *now* thinks the Danish ship must have been directed to visit Godhaab before coming here. We have left letters to go home in her, and they ought to be in England by the end of June.

I visited to-day a small lake at the foot of Mount Cunningham; it is said to occupy the centre of an extinct volcano: but I saw nothing to bear out the assertion. This is the only part of Greenland where earthquakes are felt. The Governor told me of an unusually severe shock which occurred a winter or two ago. He was sitting in his room and reading at the time, when he heard a loud noise like the discharge of a cannon; immediately afterwards a tremulous motion was felt, some glasses upon the table commenced to dance about, and papers lying upon the window-sill fell down: after a few seconds it ceased. He thinks the motion originated at the lake, as it was not felt by some people living beyond it, and that it passed from N.E. to S.W.

This mountain scenery is really charming; but a little more animal life—reindeer, for instance—would make it far more pleasing in our eyes. The last twelvemonth's produce of this district amounts only to 500 reindeer skins, instead of 3000, as in ordinary years. The clergyman of Holsteinborg was born in this colony, and has succeeded his father in the priestly office; his wife is the only European female in the colony. Being told that fuel was extremely scarce in the Danish houses, and that "the priest's wife was blue with the cold," I sent on shore a present of some coals.

On Sunday afternoon, hearing the church bell

ringing, I went on shore. It proved to be only a christening. The little dusky infant received a long string of European names. There was a small description of barrel-organ, to the sound of which the congregation joined in, keeping up a loud monotonous chant. Most of the young people had hymn-books in their hands, printed in the Esquimaux language.

Ravens seem very abundant, also large grey falcons: perhaps the dead whales may have attracted an unusual number.

Poor Christian has not only fallen desperately in love, but has engaged himself to the object of his affections, a pretty Esquimaux girl. He asked me to-day to give her a passage up to Godhavn, as he wished to leave her in charge of his mother until his return there with us next year, when his engagement for the voyage would be fulfilled. Having heard a rumour of a young woman awaiting his return with anxiety at Godhavn, I taxed him with it, but he replied with great simplicity that "he had never promised her, and would not marry her, as his friends objected to the match!" What are the good Greenlanders coming to? I recommended that he should leave his betrothed in her own home, with her mother and family. His asking a passage for her, in order to leave her with his mother, is strong proof of the sincerity of his engagement, not only to his lady love, but to the 'Fox' also.

I have written to the Admiralty to account for my prolonged absence from England; and to Dr. Rink to acquaint him with the cause of my second visit to his inspectorate.

Governor Elberg has promised to get me some fossil fish, to be found only in North Strom Fiord: they are interesting, as being of unknown geological date.

10*th*.—On the morning of the 8th we left Holsteinborg with a pleasant land wind and bright weather. When 15 miles off shore we were stopped by ice formed during the last two nights, the thermometer having fallen to 12°; out in the offing the weather was gloomy and cold, and strong northerly winds were blowing. On closing the land again, we regained the off-shore wind, and bright weather.

Keeping close alongshore, and threading our way through a vast deal of "pack" and numerous icebergs, we gained sight of Disco about noon to-day, and by the evening were within an hour's sail of Godhavn, when we were again stopped by a broad belt of ice stretching along the coast; this was a bitter disappointment, more particularly as a gale of wind with heavy sea was fast rising, and snow beginning to fall thickly; there was nothing for it, however, but to stand off under easy sail for the night.

12*th.*—At anchor at the Whalefish Islands. On the evening of the 10th we stood off from the inhospitable barrier of ice, prepared to meet the storm; snow fell so thickly that we could hardly see the icebergs in time to avoid them. We supposed ourselves to be well to leeward of the Whalefish Islands, but were deceived by the tides; suddenly a small, low islet was seen on the lee bow; not being able to pass to windward, we were obliged to wear ship, and, in doing so, passed within the ship's length of destruction—for we were certainly within that distance of the rocks! The islet was covered with snow, and, but for some very few dark points showing through, it could not be distinguished from ice. On the 11th the weather improved, and in the evening we came to our present anchorage. From a hill we can watch an opportunity to enter Godhavn. Notwithstanding the blowing weather, some natives came about five miles off to us; the water washed over their little *kayaks*, and kept the occupants' sealskin dresses streaming with wet up to their shoulders; this part of their dress seems rather part of the kayak, as it is attached to it round the hole in which the *kayaker* sits, so that no water can enter. It is wonderful to see how closely a man can assimilate his habits to those of a fish.

The Danish cooper in charge of this out-station tells us there are thirteen English whalers already out, and some of them have been up to the north end of Disco; two vessels are in sight. The world, it appears, is at peace. Petersen was at one time in charge of this station; he is now seeking out his old acquaintances.

14*th*.—Summer has suddenly burst upon us—thermometer up to 40°; moreover, we are enjoying English newspapers, and have dined off roast beef and vegetables!

Two days ago I sent a note off to a whaler by a kayak, requesting her captain to lend me some newspapers; the note reached Captain J. Walker of the 'Jane,' and yesterday his ship, accompanied by the 'Heroine,' Captain J. Simpson, approached us, and they both came in to call upon me, each of them bringing the very acceptable present of some newspapers, besides a quarter of beef, with vegetables. Nothing could exceed their sincere good feeling and kindness; they offered to supply me with anything their ships could afford. The account they give of last season is as follows: the whalers reached Devil's Point, near Melville Bay, as early as 21st May; southerly winds then set in, and blew incessantly for six weeks, during all which time they were closely beset, and the ships 'Gipsy' and 'Undaunted' were crushed. When able to move, the fleet returned southward along the "pack-edge," which was everywhere found to be impenetrable; they sailed southward of Disco, and about the middle of July the earliest ships rounded the southern extremity of the middle ice in lat. 68½°, and found no difficulty in their further passage to Pond's Bay. Captain Walker says ships could not have reached Lancaster Sound, as there was much ice north of Pond's Bay which he thought extended quite across to Melville Bay.

The position of the ice last season was considered to be most unusual; the long prevalence of southerly winds appeared to have separated the tail of the pack from the main body, the former lying against the west land about Cape Searle, whilst the latter was forced northward and pressed closely into Melville Bay; the ships sailed freely between these two great divisions, and found the west water unusually extensive.

Had I been able to collect a sufficient number

of sledge-dogs at Godhavn last year, it was my intention to have sailed across to the west side if possible, instead of pursuing the usual route through Melville Bay; but the opinions of the captains of the lost whalers were in favour of a "Melville Bay" passage, and the necessity for obtaining dogs left me no choice as to whether I should proceed west, or north to Proven and Upernivik; I have already recorded what were my opinions *at the time*, so need only observe *now*, that, although I failed, I believe my decision was justified by all former experience, even independently of the circumstances which obliged me to adopt it. Nevertheless it is mortifying to find that ships had reached as far as Pond's Bay, and with but little difficulty. Sir Edward Parry, upon his third voyage, did not reach the west water until very late in the season, although some of the whalers met with better success by following up another route.

There is nothing more uncertain than ice-navigation, dependent as it is upon winds, temperatures, and currents: one can only calculate upon "the chances," and how nearly we succeeded we have already seen. In the preceding year (1856) some of the whalers got through Melville Bay as early as the 15th June, only a few days after the commencement of the summer's thaw. Captain Walker tells me there are many years in which the whalers can pass up the western shore late in the season, but not always so far as Pond's Bay; of Melville Bay after the 10th or 15th July they know nothing, but the voyages of discovery afford us ample details; whilst of the southern route almost nothing has been made publicly known.

There are many intelligent whaling captains who possess much valuable knowledge of these lands and seas, and even in the terra incognita of Frobisher's Straits, whalers have wintered, whilst our charts scarcely afford even a vague idea of the configuration of these extensive islands. The so-called "Home Bay" has been penetrated for fifty miles, and is supposed to be a strait leading to Fox's Furthest. Scott's Inlet is also said to be a strait leading into a western arm of the same sea. A surveying vessel would be usefully employed for a couple of summers in tracing the general outline of these possessions of Her Majesty, more particularly as they are rather thickly inhabited by Esquimaux most eager to barter their produce for rifles, saws, files, knives, needles, and such like articles. Good coal has been found upon Durbin Island (near Cape Searle), in a convenient little cove upon its southern side; and as the old sailing whalers are fast being replaced by steamers, this place may become of great importance to them.

We are refitting, shooting, and devouring quantities of excellent mussels; eider ducks are very abundant, but extremely shy. Poor puss has been killed; tempted on deck by the unusually warm weather, she was pounced upon by the dogs.

17*th*.—Yesterday our attempt to enter the port of Godhavn failed, it is still filled with ice. This evening Young and I examined a narrow rocky cove—Upernivik Bay of the natives; finding it suitable for our purpose, the ship was brought in and moored to the rocks. We were received with much kindness by our friends, Mr. and Mrs. Olrik, and were presented with a file of late English papers. A considerable supply of beer was ordered to be brewed for us.

I found Mrs. Olrik without a fire in her sitting room, it was unnecessary; the windows looked to the south, and the sun shone brightly in upon a profusion of geraniums and European flowers, at once reminding one of home, and refreshing the senses by their perfume and beauty; the merry voices of the children were also a most pleasing novelty. Mr. Olrik says the past winter has not been in any way remarkable, except for the prevalence of strong winds; April and the early part of May have been unusually cold.

24*th*.—We did honour to Her Majesty's birthday by dressing the 'Fox' in all her flags, and regaling her crew with plum pudding and grog. The ice having moved off, we have come into the harbour of Godhavn, as being more convenient and safe. The day has been a busy one: we have completed our small purchases and closed our letters; I have added another Esquimaux lad to our crew, taking with him his rifle, kayak, and sledge. This evening there has been a brisk interchange of presents between us and our Danish friends. I have been given an eider-down coverlet by the Governor, Mr. Andersen, and, by Mrs. Olrik, some delicious preserve of Greenland cranberries, a tin of preserved ptarmigan, and a jar of pickled whaleskin; my table is decked with European flowers, including roses, mignonette, and violets.

With good reason shall we remember Godhavn; we have certainly been treated as especial favourites.

26*th*.—Left Godhavn early yesterday morning, and anchored this afternoon in our old position off the Coal Cliffs in the Waigat; a party of seal-hunters from Atanekerdluk came off to us, and their hunting having terminated successfully, they will assist us in coaling. From these men I obtained much information about this part of the coast; within a range of 20 miles upon the Disco shore there are four distinct coaling places; but at this early season two of them are deeply covered with snow. There is also very good coal at the S.E. end of Hare Island, where it can easily be obtained. The ice in this strait broke up as long ago as the 3rd April; it has all drifted out to the northward, only a few icebergs now remain.

28*th.*—Again hastening northward; the business of coaling was very speedily and satisfactorily completed, but the quality of the coal is very inferior. Upon the green slopes our sportsmen found nothing but a few ptarmigan and a hare.

Shortly after running close past the deserted settlement of Noursak, we arrived off a small bay, and were startled by finding the water had suddenly changed from transparent blue to a thick muddy colour, but there was no change in its depth; we were crossing the stream of "Makkaks Elvin," or Clay River, which empties itself into the bay after running through a broad and extensive valley, said to abound with reindeer; this river has its origin in lakes and glaciers in the interior, and the discolouration of the water is probably the chief cause of success in white-whale fishing, which is carried on here in the autumn, as those timid animals will not permit boats to approach them in clear water.

This evening we are crossing Omenak's Fiord, and the land-wind, which here and all along the coast northwards blows from the N.E., has come off to us.

31*st.*—Lying fast to an iceberg off Upernivik.

The whalers are all within a dozen miles of us, unable to penetrate further north. The season appears forward, and the ice much decayed; but southerly winds prevail, retarding its disruption and removal. Captain Parker, of the 'Emma,' tells me he does not expect to make a north passage this year, and as his experience extends over a period of at least thirty years, I give his reason; it is simply this,—that as during the months of February, March, and April northerly winds prevailed to an unusual degree, therefore southerly winds may now be expected to continue; if he prove a prophet, it will be to our serious hindrance at this critical season. Governor Fliescher says the winter has been mild; there has been but little wind, and that chiefly from the southward.

4th June.—We have received much kindness from our friends Captains Parker and J. Simpson, as well as from others of the whaling fleet; the former has generously supplied us with many things we were rather short of, not only in ship's stores, but provisions and coals, and in return I have of course furnished him with a receipt for his owners. Captain Simpson has most handsomely presented the 'Fox' with a sail and yards, which, after some slight alterations, will enable us to add a main topsail to our spread of canvas. For the two days we lay at the iceberg, alongside of the 'Emma,' I made furious attacks upon Captain Parker's beefsteaks and porter; we amply availed ourselves of his hearty welcome. By the arrival of the fine steam whaler 'Tay,' from Scotland, we have received papers up to 17th April.

This morning we slowly steamed away from Upernivik, threading our way betwixt islands, and ice, for about 30 miles, and now await further ice movement before it will be possible to proceed. These are called the Woman Islands, so named by the celebrated Arctic explorer John Davis, who visited them in Queen Elizabeth's reign; he found here only a few old women, their frightened lords and more active juniors having effected their escape.

Upon one of these islands a stone was picked up some 30 years ago, bearing a Runic inscription; it was sent home to Copenhagen as a most interesting relic of the early Scandinavian voyagers; but nothing was on it except the names of those men "who cleared this place" (or formed a settlement), and the date, 1135. In all probability their sojourn was extremely short, perhaps only for a single summer. The Esquimaux did not make their appearance for nearly two centuries later.

After Egede's settlement at Godhaab in 1721, the Danish trading establishments gradually extended along the coast, and Upernivik was one of them; but it appears to have been soon abandoned. During Napoleon's wars all the Danish posts were withdrawn, as the British fleet effectually cut off communication with Europe; but after peace was restored in 1815, the trading posts were again resorted to, and a new settlement formed near the ruins of the old one at Upernivik; it enjoys pre-eminence as the most northern abode of civilized man.

CHAPTER VIII.

'Fox' nearly wrecked — Afloat, and push ahead — Arctic hair-breadth escapes — Nearly caught in the pack — Shooting little auks — The Arctic Highlanders — Cape York — Crimson snow — Struggling to the westward — Reach the West-land — Off the entrance of Lancaster Sound.

June 8th.—YESTERDAY morning we passed close outside Buchan Island; it is small but lofty, its north side is almost precipitous, yet notwithstanding this strong indication of deep water, a reef of rocks lies about a mile off it. I happened to be aloft with the look-out-man at half-past eight o'clock as we were steaming through a narrow lead in the ice, when I saw a rock close ahead; it was capped with ice, therefore was hardly distinguishable from the floating masses around; the engines were stopped and reversed, but there was neither time nor room to avoid the reef, which now extended upon each side of us, and upon which the ship's bow stuck fast whilst her stern remained in 36 feet water; the tide had just commenced to fall, and all our efforts to haul off from the rocks were ineffectual. The floes lay within 30 yards of us upon each side. I feared their drifting down

THE 'FOX' ON A ROCK NEAR BUCHAN ISLAND

Drawn by Captain May

upon the ship and turning her over; but fortunately it was perfectly calm, and as the tide fell, points of the reef held them fast. The ship continued to fall over to starboard; at dead low water her inclination was 35°; the water covered the starboard gunwale from the mainmast aft, and reached almost up to the after hatchway; at this time the slightest shake must have caused her to fall over upon her side, when she would instantly have filled and sunk. The dogs, after repeated ineffectual attempts to lie upon the deck, quietly coiled themselves up upon such parts of the lee gunwale as remained above water and went to sleep.

To me the moments seemed lengthened out beyond anything I could have imagined; but at length the water began to rise, and the ship to resume her upright position. Boats, anchors, hawsers, &c., were got on board again with the utmost alacrity, and the ship floated off unhurt after having been eleven hours upon the reef. We had grounded during the day tide and were floated off by the night tide, which upon this coast occasions a much greater rise and fall,—so far we were favoured, but the poor little 'Fox' had a very narrow escape; as for ourselves, there was not the slightest cause for apprehension, three steam whalers being within signal distance.

To-day we are steaming along after the three vessels which passed us last evening and disappeared round Cape Shackleton during the night. The contrast between our prospects yesterday and to-day fills one with delight,—to be afloat and advancing unobstructedly once more is indeed charming.

11*th*.—On the afternoon of the 8th we joined the steamers 'Tay,' Captain Deuchars; 'Chase,' Captain Gravill, sen.; and 'Diana', Gravill, jun. After repeated ice-detentions, we have reached Duck Island. Captain Deuchars says, there is every prospect of an early north passage; we have had several conversations about the Pond's Bay natives, and their reports of ships, wrecks, and Europeans. There appears to be not only great difficulty, but also uncertainty, in arriving at their meaning; to form an idea of the time elapsed since an event, or the distance to the spot where it occurred, is a still harder task. I look forward to our visit at Pond's Bay with greatly increased interest.

In August, 1855, when Captain Deuchars was crossing through the middle ice, in latitude 70°, he found part of a steamer's topmast embedded in heavy ice; he also saw the moulded form of a ship's side, and thinks the latter must have sunk; the portion of the topmast visible was sawed off and taken to England. It is most probable that the vessel was either H.M.S. 'Intrepid' or 'Pioneer,' as two months later, and 250 miles further south, the 'Resolute' was picked up. About two or three years ago, Captain Deuchars lost his ship, the 'Princess Charlotte,' in Melville Bay. It was a beautiful morning; they had almost reached the North Water, and were anticipating a very successful voyage; the steward had just reported breakfast ready, when Captain Deuchars, seeing the floes closing together ahead of the ship, remained on deck to see her pass safely between them, but they closed too quickly; the vessel was *almost* through, when the points of ice caught her sides abreast of the mizenmast, and, passing through, held the wreck up for a few minutes, barely long enough for the crew to escape and save their boats! Poor Deuchars thus suddenly lost his breakfast and his ship; within *ten minutes* her royal yards disappeared beneath the surface. How closely danger besets the Arctic cruiser, yet how insidiously; everything looks so bright, so calm, so still, that it requires positive experience to convince one that ice only a very few inches, perhaps only three or four inches, *above water*, perfectly level, and moving extremely slow, could possibly endanger a strong vessel! The 'Princess Charlotte' was a very fine, strong ship, and her captain one of the most experienced Arctic seamen: he now commands the finest whaler in the fleet.

14*th.*—We have only advanced a few miles to the northward. The steamer 'Innuit' has joined our small steam squadron. Captain Sutter left Scotland only a month ago: he has very kindly and promptly sent us a present of newspapers and potatoes. Captain Deuchars has also been good enough to supply us with some potatoes and porter, perhaps the most serviceable present he could have made us after our long subsistence upon salt and preserved meats.

19*th.*—Once more alone in Melville Bay. The 'Innuit' and 'Chase' steamed much too fast for us, and the last of the four vessels, the 'Tay,' parted from us in a thick fog yesterday. We have come close along the edge of the fixed ice, passing about 6 miles outside of the Sabine Islands, and are advancing as opportunities offer. This morning the man who was stationed to watch a nip about a quarter of a mile ahead of the ship, came running back, pursued by three bears—a mother with her half-grown cubs. I suppose they followed him chiefly because he ran from them; at all events they were very close up before he reached the ship. Another bear was seen about the same time, but none of them came within shot. Rotchies (or little auks) are very abundant. Seals are occasionally shot. I ate some boiled seal to-day, and found it good: this is the first time I have eaten positive *blubber;* all scruples respecting it henceforth vanish.

25*th*.—The land-ice broke away inshore of the 'Fox' on the 19th or 20th, and we found ourselves drifting southward amongst extensive fields of ice. Sad experience has already shown us how absolutely powerless our small craft is under such circumstances. But after many attempts we regained the edge of the fast ice this morning, and steamed merrily along it towards Bushnan Island. When within a few miles a nip brought us to a standstill: here five or six icebergs lie encompassed by land-ice, and apparently aground; one of them juts out and has caught the point of an immense field of ice. There is some slight movement in the latter, but not enough to let us pass through.

Twelve or eighteen miles to the south there is a cluster of bergs, in all probability aground upon our "70 fathom bank" of last September. The ice-field appears to rest against them, as both to the east and west there is much clear water. Exactly at this spot Captain Penny was similarly detained by a nip in August, 1850. Although progress is denied to us at present, yet it is an unspeakable relief to have got out of the drifting ice.

I have passed very many anxious days in Melville Bay, but hardly any of them weighed so heavily upon me as yesterday. There was the broad, clear *land-water* within a third of a mile of me, clear weather, and a fair breeze blowing. The intervening nip worked sufficiently with wind and tide to keep one in suspense; it *nearly* opened at high water, but closed again with the ebb tide. I thought of the week already spent in struggling amongst drifting floes, and was haunted by visions of everything horrible—gales, ice-crushing, &c. Nor was it consoling to reflect that all the sailing ships as well as the steamers might have actually slipped past us. In fact, I must acknowledge that anxiety and weariness had worked me up into a state of burning impatience and of bitter chagrin at being so repeatedly baffled in all my efforts by the varying yet continual perplexities of our position. The only difference in favour of our prospects over those of the past year consisted in our having arrived here two

ESQUIMAUX IMITATING ANIMALS TO INDUCE EUROPEANS TO APPROACH.

From a sketch by Captain Allen Young.

months earlier; but the importance of this difference is incalculable.

The opportunities afforded by the delays to which we have been subjected were turned, however, to some account. Nearly one thousand rotchies were shot; they are excellent eating, their average weight is four ounces and a half, but when prepared for the table they probably do not yield more than three ounces each. A young bear imprudently swam up to the ship, and was shot,—his skin fell to the sportsman, and carcase to the dogs. Several others have been seen: we watched one fellow surprise a seal upon the ice, and carry it about in his mouth as a cat does a mouse.

27*th.*—Lying fast to the ice off the Crimson Cliffs of Sir John Ross. Yesterday we succeeded in passing through the nip, and by evening reached Cape York. Seeing natives running out upon the land-ice, the ship was made fast for an hour in order to communicate with them. A party of eight men came on board: they immediately recognised Petersen, for they lived at Etah in Smith's Sound when he was there in the American expedition. They asked for Dr. Kane, and told us Hans was married and living in Whale Sound. They all said he was most anxious to return to Greenland, but had neither sledge-dogs nor kayak; hunger had compelled him to eat the sealskin which covered the framework of the latter. Petersen gave them messages for Hans from his Greenland friends, and advice that he should fix his residence here, where he might see the whalers and perhaps be taken back to Greenland. The natives did not seem to be badly off for anything except dogs, some distemper having carried off most of these indispensable animals, I was therefore unable to procure any from them. These people spent the winter here; they seemed healthy, well-clad, and happy little fellows. One of them is brother-in-law to Erasmus York, who voluntarily came to England in the 'Assistance' in 1851. This man is an *angekok*, or magician; he has a still flatter face than the rest of his countrymen, but appears more thoughtful and intelligent.

Petersen pointed out to me a stout old fellow, with a tolerable sprinkling of beard and moustache. This worthy perpetrated the only murder which has taken place for several years in the tribe: he disliked his victim and stood in need of his dogs, therefore he killed the owner and appropriated his property! Such motives and passions usually govern the "unsophisticated children of nature;" yet, as savages, the

Esquimaux may be considered exceedingly harmless.

Of late years these Arctic Highlanders have become alarmed by the rapid diminution of their numbers through famine and disease, and have been less violent towards each other in their feuds aud quarrels.

The appearance of these men, as they danced and rolled about in frantic delight at our approach, was wild and strange, and their costume uniform and picturesque. Their long, coarse, black hair hung loosely over the sealskin frock, which in its turn overlapped their loose shaggy bearskin breeches, and these again came down over the tops of their sealskin boots. Most of them carried a spear formed out of the horn of a narwhal.

Having distributed presents of knives and needles, and explained to them that we did so because they had behaved well to the white people (as we learn from Dr. Kane's narrative of their treatment of him and his crew), we pursued our voyage, not doubting but that we should soon reach the *North Water*, an extensive sea through which we could sail uninterruptedly to Pond's Bay.

During the night we advanced through loose ice; but fog and a rising S.E. gale delayed us, and to-day the pack has pressed in against the land, so that our wings are most unexpectedly clipped. A walrus was shot through the head by a Minié bullet; none other will penetrate such a massive skull: unfortunately for my collection of specimens and for the dogs, the animal sank.

2nd July.—For five days we have been almost beset amongst loose ice and grounded bergs; the winds were generally from the S.E. and accompanied by fog. To avoid being squeezed we had constantly to shift our position; once we were caught and rather severely nipped; the ship was heeled over about ten degrees and lifted a couple of feet: the ice was three feet thick, but broke readily under her weight. Unfortunately there was not time to unship the rudder, so it suffered very severely. Upon a previous occasion the screw-shaft was bent and a portion of the screw broken off.

Landed to obtain a good view of the sea in the offing; from the hills we could see nothing but pack to seaward. There was no land ice; we stepped out of the boat upon a narrow ice-foot which fringed the coast; immediately above it we trod over a velvet sward of soft bright-green moss; the turf beneath was of considerable depth. Here and there under this noble range of cliffs, which are composed of primary rock, there exists much vegetation for so high a latitude. From the fact of thick layers of turf descending quite down to the sea, it is evident that the land has been gradually sinking. Steep slopes of rocky *débris*, which screen the bases of the most precipitous cliffs, form secure nurseries for the little auk; these localities were literally alive with them; they popped in and out of every crevice, or sat in groups of dozens upon every large rock. I have nowhere seen such countless myriads of birds. The *rotchie*, or little auk, lays its single egg upon the bare rock, far within a crevice beyond the reach of fox, owl, or burgomaster gull. We shot a couple of hundred during our short stay on shore, and, by removing the stones, gathered several dozen of their eggs.

The huge predatory gulls, long ago named "Burgomasters" by the Dutch seamen (because they lord it over their neighbours, and appropriate everything good to themselves), have established themselves in the cliffs, where their nests are generally inaccessible: we were a month too late for their eggs; the young birds were as large as spring chickens. Of course we obtained specimens of the red snow, but had to seek rather diligently for it; its colour was a dirty red, very like the stain of port wine: very few patches of it were found.

Last night a westerly wind blew freshly and dispersed the ice outside of us, so much so that this evening we have got out into almost clear water. Farewell, Greenland,—hurrah for the west!

5*th.*—After getting free from the ice off the Crimson Cliffs, we soon lost sight of the last fragment, and steered for Pond's Bay. And now we all set to work in zealous haste to write our last letters for England, by the whalers, which we hoped soon to meet there.

After running 60 miles the ice reappeared, and we sailed through a vast deal of it, but it became more closely packed, and a thick fog detained us for a day.

When the weather became clear, the main pack was seen to the W., S., and S.E.; in the hope of rounding its northern extreme we ran along it to the N.W. To-day it has led us to the N. and N.E., so that this evening Wolstenholme Sound is in sight. To the N. the pack appears impenetrable, and there is a strong ice-blink over it. All the ice we have lately sailed through is loose, and much decayed; it seems but recently to have broken away from the land, is not water-washed, neither has it been exposed to a swell, the fractured edges remaining sharp.

6th.—Midnight. Last evening I persevered to the N. until every hope of progress in that direction vanished. To the W. the pack appeared tolerably loose; the wind was fresh at E.S.E., so I determined once more to push into it, and endeavour to battle our way through; I hoped it would prove to be merely a belt of 30 or 40 miles in width. We found the ice to lie for the most part in streams at right angles to the wind, and therefore much more open than it had appeared: there was seldom any difficulty in winding through it from one water space to another. The wind greatly increased, bringing much rain, but fortunately no fog;—the dread of this hung over me like a nightmare,—our progress depended upon the vigilance of the look-out kept in the crow's-nest. By noon we had made good 60 miles. Throughout the day the wind has gradually moderated; the rain gave place to snow, which in its turn was succeeded by mist. The evening was fine eventually and clear; but still we find the ice is all around. Just before midnight the termination of our lead was discovered, whilst the ice through which we had passed was closing together, and a dense fog came rolling down. Under these circumstances the ship was made fast as near to the nip as safety permitted, to await some favourable change.

10*th.*—All the 7th we remained in our small basin, there being no outlet from it, and but little water anywhere visible. To pass away the dull hours and get rid of unwelcome reflections upon the similarity of our present position and that in August last, I commenced an attack upon all the feathered denizens of the pack—they seemed so provokingly contented with it—but they soon became wary, and deserted our vicinity, so I shot only a dozen fulmar petrels, three ivory gulls, two looms,* and a *Lestris parasiticus;* some of them were useful as specimens, and such as were not destined for our table were given to the dogs. Although Cobourg Island was 45 miles distant from us, its lofty rounded outlines were very distinct, and much covered with snow. On the 8th we squeezed through nips for 4 or 5 miles, and on the 9th, reaching a large space of water, steamed towards Cobourg Island until again stopped by the pack at an early hour this morning, when within 5 or 6 leagues of it.

This evening we are endeavouring to steam

* These birds are called willocks at home; they are the "Uria Brunnichii" of naturalists.

in towards the West-land, and fancy we can trace with the crow's-nest telescope a practicable route through the intervening ice-mazes to a faint streak of water along the shore. This sort of navigation is not only anxious, but wearying. To me it seems as if several months instead of only eight days had elapsed since we left Cape York. We are constantly wondering what our whaling friends are about, and where they are?

14*th*.—The faint streak of water seen on the night of the 10th proved to be an extensive sheet to leeward of Cobourg Island. We reached it next morning. Jones' Sound appeared open, and a slight swell reached us from it, but all along the shore there was close pack. Although but little water was visible to the southward, we persevered in that direction, and, as the ice was rapidly moving offshore under the combined influence of wind and tide, we were only occasionally detained.

Two hundred and forty-two years ago—to a day, I believe—William Baffin sailed without hindrance along this coast and discovered Lancaster Sound. What a very different season he must have experienced!

Passing near Cape Horsburgh we approached De Ros Islet at midnight. The air being very calm, and still, the shouting of some natives was heard, although we could scarcely distinguish them upon the land-ice. The ship was made fast, and the shouting party, consisting of three men, three women, and two children, eagerly came on board. Only four individuals remained on shore.

The old chief Kal-lek is remarkable amongst Esquimaux for having a bald head. He inquired by name for his friend Captain Inglefield. These three families have spent the last two years upon this coast, between Cape Horsburgh and Croker Bay. Their knowledge does not extend further in either direction. They are natives of more southern lands, and crossed the ice in Lancaster Sound with dog-sledges. Since the visit of the 'Phœnix' in '54 they have seen no ships, nor have any wrecks drifted upon their shores. They seemed very fat and healthy, but complained that all the reindeer had gone away, and asked if *we* could tell where they went to? Our presents of wood, knives, and needles were eagerly received. They assured us that Lancaster Sound was still frozen over, and that all the sea was covered with pack. After half an hour's delay we steamed onward, and on reaching a larger space of water our hopes (somewhat depressed by the native intelligence) began to revive. But we soon found that our clear water terminated near Cape Warrender. Lancaster Sound, although not frozen over, was crammed full of floes and icebergs. The wind increased to a strong gale from the east, and pressed in more ice. At length the ship was with difficulty made fast to a strip of land-ice a few miles westward of Point Osborn. Gradually the gale subsided, but not until the pack was close in against the land. The tides kept sweeping it to and fro, to our great discomfort. The land is composed of gneiss, and the gravelly shore is low. A few ducks only have been shot, and traces of reindeer and hares seen. Our Melville Bay friends, the rotchies, are very rare visitors upon this side of Baffin's Bay.

Part of a ship's timber has been found upon the beach; it measures 7 inches by 8 inches, is of American oak, and, although sound, has long been exposed to the weather.

CHAPTER IX.

Off Cape Warrender — Sight the whalers again — Enter Pond's Bay — Communicate with Esquimaux — Ascend Pond's Inlet — Esquimaux information — Arctic summer abode — An Arctic village — No intelligence of Franklin's ships — Arctic trading — Geographical information of natives — Information of Rae's visit — Improvidence of Esquimax — Travels of Esquimaux.

6th July.—To borrow a whaling phrase, we are "dodging about in a hole of water" off Cape Warrender. I recognise the little bay just to the west of the cape where Parry landed in September, 1824. The "immense mass of snow and ice containing strata of muddy-looking soil" is there still, and, I should think, had considerably increased. Here his party shot three reindeer out of a small herd. We have narrowly scanned the steep hill-sides with our glasses, but without discovering any such inducement to land.

No cairns are visible upon Cape Warrender; the natives have probably removed them. Dense pack prevents us from approaching Port Dundas or crossing to the southern shore. We all find these vexatious delays are by no means conducive to sleep. The mind is busy with a sort of magic-lantern representation of the past, the present, and the future, and resists for weary hours the necessary repose.

17*th.*—Last night's calm has allowed the pack to expand so much, that to-day we have steamed through it until within three miles of the noble cliffs of Cape Hay; and now we are drifting eastward with the ice precisely as did the 'Enterprise' and 'Investigator' in September, '49. Upon that occasion we were set free off Pond's Bay. There is a very extensive *loomery* at Cape Hay; we regret the circumstances which prevent our levying a tax upon it. Here, if anywhere, I expected to find a clear sea, but east winds have prevailed for twenty days out of the last twenty-five, and this accounts for the present state of the sea; the next succession of west winds will probably effect a prodigious clearance of ice.

21*st.*—The 'Tay' was seen to-day in loose ice, and much further off the land. She gradually steamed through it to the southward, and by night was almost out of sight. Her appearance surprised us, as we supposed she must have reached Pond's Bay long ago. Ten hours' struggling with steam and sails at the most favourable intervals has only advanced us five miles. The weather is remarkably warm, bright, and pleasant. A very large bear came within 150 yards, and was shot by Petersen, the Minié bullet passing through his body. This beast measured 8 ft. 3 in. in length; his fat carcase was hoisted on board with great satisfaction, as our dogs' food was nearly expended.

24th.—Last night the ice became slack enough to afford some prospect of release, so we charged the nips vigorously, and steamed away through devious openings towards Cape Fanshawe. For several hours but little progress was made, but this morning the ice became more open; clear water was seen ahead, and reached by noon. Although it is calm I prefer waiting for a breeze to expending more coals. We are only ten miles from Possession Bay. The air is so very clear that the land appears quite close to us. All that is not mountainous is well cleared of snow. There is immense refraction. Only a single iceberg in sight. The sea-water is light green, as remarked by Parry in 1819.

26th.—A vessel was seen yesterday morning; the day continuing calm, we steamed through some loose ice, and joined her off Cape Walter Bathurst in the evening. It proved to be the 'Diana;' she parted from us on the 16th of June in Melville Bay, has everywhere been obstructed by the pack, as we have been, and only reached Cape Warrender three days before us. From thence to Possession Bay she met with *no obstruction.* The subsequent east winds brought in all the ice which has so much retarded us.

The 'Diana' has already captured twelve whales. Taking the hint from Capt. Gravill, we have made fast to a loose floe, and are drifting very nearly a mile an hour to the southward along the edge of very formidable land-ice, which is seven or eight miles broad. All to seaward of us is packed ice. The old whaling seamen of the 'Diana' are astounded at the unusual and unaccountable abundance of ice which everywhere fills up Baffin's Bay. All the 'Diana's' steaming-coals, her spare spars, wood, and even a boat, have been burnt in the protracted struggle through the middle ice.

27*th*.—After putting our letter-bag on board the 'Diana' this morning we steamed on for Pond's Bay, and at noon made fast near Button Point to the land-ice, which still extends across it.

For four hours Petersen and I have been bargaining with an old woman and a boy, not for the sake of their seal-skins, but in order to keep them in good humour whilst we extracted information from them. They said they knew nothing of ships or white people ever having been within this inlet, nor of any wrecked ships. They knew of the depôt of provisions left at Navy Board Inlet by the 'North Star,' but had none of them. The woman has traced on paper the shores of the inlet as far as her knowledge extends, and has given me the name of every point. She says the ice will break up with the first fresh wind. These two individuals are alone here. They remained on purpose to barter with the whalers, and cannot now rejoin their friends, who are only 25 miles up the inlet, because the ice is unsafe to travel over and the land precipitous and impracticable.

This afternoon the 'Tay' stood in towards us, and Captain Deuchars kindly sent his boat on board with an offer to take charge of our letters. The 'Tay' reached this coast only a few days ago, having met with the same difficulties which we experienced. The 'Innuit' was last seen nearly a month ago beset off Jones' Sound. The remaining steamer, the 'Chase,' has not been seen or heard of.

29*th*.—The old woman's denial of all knowledge of wrecks or cast-away men was very unsatisfactory. I determined to visit her countrymen at their summer village of Kaparōktolik, which she described as being only a short day's journey up the inlet.

Petersen and one man accompanied me. We started yesterday morning with a sledge and a Halkett boat. Although the ice over which we purposed travelling broke away from the land soon after setting out, yet we managed to get half way to the village before encamping. This morning we learnt the truth of the old woman's account. A range of precipitous cliffs rising from the sea cut us off by land from Kaparōktolik, so we were obliged to return to the ship. Our walk afforded the opportunity of examining some native encampments and câches. We found innumerable scraps of seal-skins, bird-skins, walrus and other bones, whalebone, blubber, and a small sledge. The latter was very old, and composed of pieces of wood and of large bones ingeniously secured together with strips of whalebone. Five preserved-meat tins were found; some of them retaining their original coating of red paint. Doubtless these were part of the spoils from Navy Board Inlet depôt. The total absence of fresh wood or iron was strongly in favour of the old woman's veracity. Since yesterday, ice, about 16 miles in extent, has broken up in the inlet, and is drifting out into Baffin's Bay.

During my absence our shooting parties have twice visited a *loomery* upon Cape Graham

Moore, and each time have brought on board 300 looms. Very few birds and no other animals were seen during our walk over the rich mossy slopes to-day. I saw a pair of Canadian brown cranes, the first of the species I have ever seen so far north, though Sir Robert M'Clure found them, I know, on Bank's Land.

The lands enjoying a southern aspect, even to the summits of hills 700 or 800 feet in height, were tinged with green; but these hills were protected by a still loftier range to the north. Upon many well-sheltered slopes we found much rich grass. All the little plants were in full flower; some of them familiar to us at home, such as the buttercup, sorrel, and dandelion. I have never found the latter to the north of 69° before.

The old woman is much less excited to-day; she says there was a wreck upon the coast when she was a little girl; it lies a day and a half's journey, about 45 miles, to the north; and came there without masts and very much crushed; the little which now remains is almost buried in the sand. A piece of this wreck was found near her *abode*,—she has neither hut nor tent, but a sort of lair constructed of a few stones and a seal-skin spread over them, so that she can crawl underneath. This fragment is part of a floor timber, English oak, 7½ inches thick; it has been brought on board.

30*th*.—A gale of wind and deluge of rain has detained the ship until this evening; we are now steaming up the inlet, having the old lady and the boy on board as our pilots; they are delighted at the prospect of rejoining their friends, from whom they were effectually cut off until the return of winter should freeze a safe pathway for them; they had, however, abundance of looms stored up *en câche* for their subsistence. She has drawn me another chart, much more neatly than the former, but so like it as to prove that her geographical knowledge, and not her powers of invention, have been taxed. She is a widow; her daughter is married, and lives at a place called Igloolik, which is six or seven days' journey from here,—three days up the inlet, then about three days overland to the southward, and then a day over the ice.

Thinking it not quite impossible that this Igloolik might be the place where Parry wintered in 1822-3, I told Petersen to ask whether ships had ever been there? She answered, "Yes, a ship stopped there all one winter; but it is a long time ago." All she could distinctly recollect having been told about it was, that one of the crew died, and was buried there, and his name was Al-lah or El-leh. On referring to Parry's 'Narrative,' I found that the ice-mate, Mr. Elder, died at Igloolik! This is a very remarkable confirmation of the locality,—for there are several places called Igloolik. She also told us it was an island, and near a strait between two seas. The Esquimaux take considerable pains to learn, and remember names; this woman knows the names of several of the whaling captains, and the old chief at De Ros Islet remembered Captain Inglefield's name, and tried hard to pronounce mine.

She now told us of another wreck upon the coast, but many days' journey to the south of Pond's Bay; it came there before her first child was born. Her age is not less than forty-five.

August 4th.—Our Esquimaux friends have departed from us with every demonstration of friendship, to return to their village. We have had free communication with them for four days—not only through Mr. Petersen, but also through our two Greenlanders; the result is, that they have no knowledge whatever of either the missing or the abandoned searching ships. Neither wrecked people nor wrecked ships have reached their shores. They seemed to be much in want of wood; most of what they have consists of staves of casks, probably from the Navy Board Inlet depôt.

In their bartering with us, saws were most eagerly sought for in exchange for narwhal's horns; they are used by them in cutting up the long strips of the bones of whales with which they shoe the runners of their sledges, also the ivory and bone used to protect the more exposed parts of their kayaks and the edges of their paddles from the ice.

Files were also in great demand, and I found were required to convert pieces of iron-hoop into arrow and spear-heads. If any suspicion existed of their having a secret supply of wood such as a wreck or even a boat would afford, it was removed by their refusing to barter the most trifling things for axes or hatchets.

But I must relate the events of the last few days as they occurred. When 17 miles within the inlet we reached the unbroken ice and made the ship fast. Here the *strait*—originally named Pond's *Bay*, and more recently Eclipse *Sound*—appears to be most contracted, its width not exceeding 7 or 8 miles. Both its shores are very bold and lofty, often forming noble precipices. The prevailing rock is grey gneiss, generally dipping at an angle of 35° to the west.

Early on the 1st of August I set out for the native village with Hobson, Petersen, two men, and the two natives from Button Point. Eight miles of wet and weary ice-travelling, which occupied as many hours, terminated our journey; the surface of the ice was everywhere deeply channelled, and abundantly flooded by the summer's thaw: we were almost constantly launching our small boat over the slippery ridges which separated pools or channellings through which it was generally necessary to wade.

After toiling round the base of a precipice, we came rather suddenly in view of a small semicircular bay; the cliffs on either side were 800 or 900 feet high, remarkably forbidding and desolate; the mouth of a valley or wide mountain gorge opens out into its head. Here, in the depth of the bay, upon a low flat strip of land, stood seven tents,—the summer village of Kaparōk-to-lik. I never saw a locality more characteristic of the Esquimaux than that which they have here selected for their abode;—it is wildly picturesque in the true Arctic application of the term.

THE VILLAGE AND GLACIER OF KAPAROKTOLIK, GREENLAND.

Drawn by Captain May.

Although August had arrived, and the summer had been a warm one, the bay was still frozen over; and if there was an ice-covered *sea* in front, there was also abundance of ice-covered *land* in the rear—a glacier occupied the whole valley behind, and to within 300 yards of the chosen spot!

The glacier's height appeared to be from 150 to 200 feet; its sea-face extending across the valley,—a probable width of 300 or 400 yards,—was quite perpendicular, and fully 100 feet high. All last winter's snow had thawed away from off it and exposed a surface of mud and stones, fissured by innumerable small rivulets, which threw themselves over the glacier cliffs in pretty cascades, or shot far out in strong jets from their deeply serried channels in its face; whilst other streamlets near the base burst out through sub-glacial tunnels of their own forming.

What a strange people to confine themselves to such a mere strip of beach! Upon each side they have towering rocky hills rising so abruptly from the sea, that to pass along their bases or ascend over their summits, is equally impossible; whilst a threatening glacier immediately behind, bears onward a sufficient amount of rock and earth from the mountains whence it issues, to convince even the unreflecting savage of its progressive motion.

The land is devoid of game, although lemmings and ermines are tolerably numerous; it only supplies the moss which the natives burn with blubber in their lamps, and the dry grass which they put in their boots; even the soft stone, *lapis ollaris*, out of which their lamps and cooking vessels are made, and the iron pyrites with which they strike fire, are obtained by barter from the people inhabiting the land to the west of Navy Board Inlet. But the sea compensates for every deficiency. The assembled population amounted to only 25 souls: 9 men, the rest women and children.

All of them evinced extreme delight at seeing us; as we approached the huts the women and children held up their arms in the air and shouted "Pilletay" (give me), incessantly; the men were more quiet and dignified, yet lost no opportunity, either when we declined to barter, or when they had performed any little service, to repeat "Pilletay" in a beseeching tone of voice.

We walked everywhere about the tents and entered some of them, carefully examining every chip or piece of metal; our visit was quite unexpected. They had only two sledges; both were made of 2½-inch oak-planks, devoid of bolt-holes or treenails, and having but very few nail-holes. These sledges had evidently been constructed for several years, the parts not exposed to friction were covered with green fungus: one of them measured 14 feet long, the other about 9 feet; we were told the wood came from a wreck to the southward of Pond's Bay. Most of the sledge crossbars were ordinary staves of casks. Amongst the poles and large bones which supported the tents we noticed a painted fir oar. Some pieces of iron-hoop and a few preserved-meat tins—one of which was stamped "Goldner,"—completed their stock of European articles.

Petersen questioned all the men *separately* as to their knowledge of ships or wrecks; but their accounts only served to confirm the old woman's story. None of them had ever heard of ships or wrecks anywhere to the westward. Both individually and collectively we got them to draw charts of the various coasts known to them, and to mark upon them the positions of the wrecks. The two chiefs, Nōo-luk and Ā-wăh-lah, soon made themselves known to me, and, when we desired to go to sleep, sent away the people who were eagerly pressing round our tent. All these natives were better-looking, cleaner, and more robust than I expected to find them.

A-wăh-lah has been to Igloolik; one of his wives, for each chief has *two*, has a brother living there. I spread a large roll of paper upon a rock, and got him to draw the route overland, and also round by the coast to it; this novel proceeding attracted the whole population about us; A-wăh-lah constantly referred to others when his memory failed him; at length it was completed to the satisfaction of all parties. When I gave him the knife I had promised as his reward, and added another for his wives, he sprang up on the rock, flourished the knives in his hands, shouted, and danced with extravagant demonstrations of joy. He is a very fine specimen of his race, powerful, impulsive, full of energy and animal spirits, and moreover an admirable mimic. The men were all about the same height, 5 feet 5 in.; they eagerly answered our questions, and imparted to us all their geographical knowledge, although at first they hesitated when we asked them about Navy Board Inlet, in consequence of the depôt placed there having been plundered; but we soon found that they were easily tired under cross-examination, and often said they knew no more; it was necessary to humour them.

According to their account the depôt was discovered and robbed by people living further west. This is probably true, as so few relics were to be seen here, which would not be the case if such active fellows as A-wăh-lah and Nōo-luk had received the first information of its proximity. These people of Kaparōktolik are the only inhabitants of the land lying eastward of Navy Board Inlet, and live entirely upon its *southern* shore. In a similar manner, it is only the *southern* coast of the land to the west of Navy Board Inlet that is inhabited. After distributing presents to all the women and children, and making a few trifling purchases from the men, we returned next day to the ship.

During my absence more ice had broken away, involving the ship and almost forcing her on shore. It required every exertion to save her. For two hours she continued in imminent danger, and was only saved by the warping and ice-blasting, by which at last she got clear of the drifting masses, *four minutes* only before these were crushed up against the rocks!

Four Esquimaux came off to the ship in their

kayaks, bringing whalebone, narwhals' horns, &c., to barter. Next to handsaws and files, they attached the greatest value to knives and large needles. These men remained on board for nearly two days, and drew several charts for us. Nōo-luk explained that seven or eight days' journey to the southward there are *two* wrecks a short day's journey apart. The southern is in an inlet or strait which contains several islands, but here his knowledge of the coast terminates. The man A-ra-neet said he visited these wrecks five winters ago. All of them agreed that it is a very long time since the wrecks arrived upon the coast; and Nōo-luk, who appears to be about forty-five years of age, showed us how tall he was at the time.

In the 'Narrative of Parry's Second Voyage,' at p. 437, mention is made of the arrival at Igloolik of a sledge constructed of ship-timber and staves of casks; also of two ships that had been driven on shore, and the crews of which went away in boats. In August, 1821, nearly two years previous to the arrival of this report through the Esquimaux to Igloolik, the whalers 'Dexterity' and 'Aurora' were wrecked upon the west coast of Davis' Strait, in lat. 72°, 70 or 80 miles southward of Pond's Bay. The old man, Ow-wang-noot, drew the coast-line northwards from Cape Graham Moore to Navy Board Inlet, and pointed out the position of the northern wreck a few miles east of Cape Hay. Had it been conspicuous, we must have seen it when we slowly drifted along that coast.

These people usually winter in snow-huts at Green Point, a mile or two within the northern entrance of Pond's Bay. They hunt the seal and narwhal, but when the sea becomes too open they retire to Kaparōktolik; and when the remaining ice breaks up—usually about the middle of August—a further migration takes place across the inlet to the S.W., where reindeer abound, and large salmon are numerous in the rivers. Every winter they communicate with the Igloolik people. Two winters ago (1856-7) some people who live far beyond Igloolik, in a country called A-ka-nee (probably the Ak-koo-lee of Parry), brought from there the information of white people having come in two boats, and passed a winter in snow-huts at a place called by the following names:—A-mee-lee-oke, A-wee-lik, Net-tee-lik.

Our friends pointed to our whale-boat, and said the boats of the white people were like it, but larger. These whites had tents inside their snow-huts; they killed and eat reindeer and narwhal, and smoked pipes; they bought dresses from the natives; none died; in summer they all went away, taking with them two natives, a father and his son. We could not ascertain the name of the white chief, nor the interval of time since they wintered amongst the Esquimaux, as our friends could not recollect these particulars.*

The name of the locality, A-wee-lik (spelt as written down at the moment), may be considered identical with "Aÿ-wee-lik," the name of the land about Repulse Bay in the chart of the Esquimaux woman, Iligliuk (Parry's 'Second Voyage,' p. 197).

We were of course greatly surprised to find that Dr. Rae's visit to Repulse Bay was known to this distant tribe; and also disappointed to find they had heard nothing of Franklin's Back-River parties through the same channel of communication. They were anxiously and repeatedly questioned, but evidently had not heard of any other white people to the westward, nor of their having perished there.

Ow-wang-noot lived at Igloolik in his early days, and made a chart of the lands adjacent, but said he was so young at the time that "it seemed like a dream to him." He was ac-

* Dr. Rae wintered at Repulse Bay in *stone* huts in 1846-7. Again wintered there in *snow* huts in 1853-4.

quainted with Ee-noo-lōō-apik, the Esquimaux who once accompanied Captain Penny to Aberdeen, and told us he had died, lately I think, at a place to the southward called Kri-merk-sū-malek, but that his sister still lives at Igloolik.

Although they told us the Igloolik people were worse off for wood than they were themselves, yet it was evident that here also it is very scarce. We could not spare them light poles or oars such as they were most desirous to obtain for harpoon and lance staves and tent-poles; and they would willingly have bartered their kayaks to us for rifles (having already obtained some from the whaling-ships), but that they had no other means of getting back to their homes, nor wood to make the light framework of others.

They collect whalebone and narwhals' horns in sufficient quantity to carry on a small barter with the whalers. A-wăh-lah showed us about thirty horns in his tent, and said he had many more at other stations. A few years ago, when first this bartering sprang up, an Esquimaux took such a fancy to a fiddle that he offered a large quantity of whalebone in exchange for it. The bargain was soon made, and subsequently this whalebone was sold for upwards of a hundred pounds! Each successive year, when the same ship returns to Pond's Bay, this native comes on board to visit his friends, and goes on shore with many presents in remembrance of the memorable transaction. It is much better for him thus to receive annual gifts than to have received a large quantity at first, as the improvidence of these men surpasses belief.

Of the "rod of iron about four feet long, supposed to have been at one time galvanised," which was brought home in 1856 by Captain Patterson, and forwarded to the Admiralty, I could obtain no information. The natives were shown galvanised iron, and said they had never seen any before; if their countrymen had any, it must have come from the whalers; none like it was found in the wrecks. Rod-iron is very valuable to Esquimaux for spears and lances, and narwhals' horns very tempting to the seamen, not only as valuable curiosities, but the ivory is worth half-a-crown a pound; and I have but little doubt that many of the things said to have been stolen by the natives were fraudulently bartered away by the sailors. That there was no galvanised iron on board any of the Government searching-ships, nor in the missing expedition which sailed from England as far back as 1845, I am almost certain. But is it *certain* that this iron rod was galvanised?

The natives gave Captain Patterson to understand that they got it from the wreck to the north.

In July, 1854, Captain Deuchars was at Pond's Bay, and many natives visited his ship, coming over the ice on twelve or fourteen sledges made of ship's planking. Now at this time Sir Edward Belcher's ships were still frozen up in Barrow Strait. My own impression is that the natives whom Captain Deuchars communicated with in 1854 were visitors at Pond's Bay—certainly from the *southward*—and probably attracted by the barter recently grown up at that whaling rendezvous. Having discovered the use of the saws obtained by barter from our whalers, they had successfully applied them to the stout planking of the old wrecks, which they could not have stripped off with any tools previously in their possession.

That the various tribes, or rather groups of families, occasionally visit each other, sometimes for change of hunting-grounds, but more frequently for barter, is well known. Captain Parker told me that a native whom he had met one summer at Durbin Island, came on board his ship at Pond's Bay the following year. The distance between the two places, as travelled by this man in a single winter, is scarcely short of

500 miles; and the information given us of Rae's wintering at Repulse Bay, information which must have travelled here in two winters, shows that these natives communicate at still greater distances.

Did other wrecks exist nearer at hand, our Pond's Bay friends would be much better supplied with wood. If the Esquimaux knew of any within 300, 400, or even 500 miles, the Pond's Bay natives would at least have heard of them, and could have had no reason for concealing it from us. I only regret that we had not the good fortune to see more than a few natives, and but two sledges of ship's planking; otherwise our own information might have been more copious, and the origin of the fresh supply of planking decisively ascertained.

CHAPTER X.

Leave Pond's Bay — A gale in Lancaster Sound — The Beechey Island depôt — An Arctic monument — Reflections at Beechey Island — Proceed up Barrow's Strait — Peel Sound — Port Leopold — Prince Regent's Inlet — Bellot Strait — Flood-tide from the west — Unsuccessful efforts — Fox's Hole — No water to the west — Precautionary measures — Fourth attempt to pass through.

6th Aug.—Continued calms have delayed us. This evening we steamed from Pond's Bay northward, although our coals have been sadly reduced by the almost constant necessity for steam-power since leaving the Waigat. The three steam-whalers have gone southward; none others have arrived. They appear to us to be leaving the whales behind them; we saw many whilst up the strait, and at the edge of the remaining ice. The natives said they would remain as long as the ice remained, but when it all broke up they would return into Baffin's Bay and go southward; and that these animals arrive in early spring, and do not pass through the strait into any other sea beyond.

Monday evening, 9th.—On the night of the 6th a pleasant fair breeze sprang up, and enabled us to dispense with the engine. An immense bear was shot; he measured 8 feet 7 inches in length, and is destined for the museum of the Royal Dublin Society. On the 7th the wind gradually freshened and frustrated my intention of examining the wreck spoken of near Cape Hay; at night it increased to a very heavy gale. Although past Navy Board Inlet, very little ice had yet been met with. The weather, and fear of ice to leeward, obliged us to heave the vessel to, under main trysail and fore staysail. The squalls were extremely violent and seas unusually high.

All Sunday, the 8th, the gale continued, although not with such extreme force; the deep rolling of the ship, and moaning of the half-drowned dogs amidst the pelting sleet and rain, was anything but agreeable. Notwithstanding that I had been up all the previous night, I felt too anxious to sleep; the wind blew directly up Barrow Strait, drifting us about two miles an hour. Occasionally she drifted to leeward of masses of ice, reminding us that if any of the dense pack which covered this sea only three weeks ago remained to leeward of us, we must be rapidly setting down upon its weather edge. The only expedient in such a case is to endeavour to run into it—once well within its outer margin a ship is comparatively safe—the danger

THE 'FOX' ARRIVING AT BEECHEY ISLAND

Drawn by Captain May.

lies in the attempt to penetrate; to escape out of the pack afterwards is also a doubtful matter.

In the evening we were glad to see the land, and find ourselves off the north shore near Cape Bullen, for the violent motion of the ship and very weak horizontal magnetic force had rendered our compasses useless. This morning, the 9th, the gale broke, and the sea began to subside rapidly; by noon it was almost calm, but a thick gloom prevailed, ominous, it might be, of more mischief. All along the land there is ice, but broken up into harmless atoms. We have carried away a maingaff and a jibstay, but have come remarkably well through such a gale with such trifling damage.

11*th.*—Before noon to-day we anchored inside Cape Riley, and immediately commenced preparations for embarking coals. I visited Beechey Island house, and found the door open; it must have been blown in by an easterly gale long ago, for much ice had accumulated immediately inside it. Most of the biscuit in bags was damaged, but everything else was in perfect order. Upon the north and west sides of the house, where a wall had been constructed, there was a vast accumulation of ice, in which the lower tier of casks between the two were embedded, and its surface thawed into pools. Neither casks nor walls should have been allowed to stand near the house. The southern and eastern sides were clear and perfectly dry. The 'Mary' decked boat, and two 30-feet lifeboats, were in excellent order, and their paint appeared fresh, but oars and bare wood were bleached white.

The gutta-percha boat was useless when left here, and remains in the same state. Two small sledge travelling boats were damaged; one of them had been blown over and over along the beach until finally arrested by the other. The bears and foxes do not appear to have touched anything. I have taken on board all letters left here for Franklin's or Collinson's expeditions, and also a 20-feet sledge-boat for our own travelling purposes.

Last night we steamed very close round Cape Hurd in a dense fog, and crept along the land as our only guide: we were thus led into Rigby Bay, and discovered a shoal off its entrance by grounding upon it. After a quarter of an hour we floated off unhurt.

In lowering a boat to pursue a bear, Robert Hampton fell overboard; fortunately he could swim, and was very soon picked up, but the intense cold of the water had almost paralyzed his limbs. The bear was shot and taken on board.

Sunday, 15*th*, 9 P.M.—Our coaling was completed yesterday, and the ship brought over and anchored off the house in Erebus and Terror Bay. A small proportion of provisions and winter clothing has been embarked to complete our deficiencies; the ice has been scraped out of the house and its roof thoroughly repaired, a record deposited, and door securely closed.

I found lying at Godhavn a marble tablet which had been sent out by Lady Franklin, in the American expedition of 1855 under Captain Hartstein, for the purpose of being erected at Beechey Island. Circumstances prevented the Americans executing this kindly service, and it fell to my lot to convey it to the site originally intended. The tablet was constructed in New York under the direction of Mr. Grinnell at the request of Lady Franklin, in order that the only opportunity which then offered of sending it to the Arctic regions might not be lost. I placed the monument upon the raised flagged square in the centre of which stands the cenotaph recording the names of those who perished in the Government expedition under Sir Edward Belcher. Here also is placed a small tablet to the memory of Lieutenant Bellot. I could not have selected for Lady Franklin's memorial a more appropriate or conspicuous site. The inscription runs as follows:—

TO THE MEMORY OF

FRANKLIN,

CROZIER, FITZJAMES,

AND ALL THEIR
GALLANT BROTHER OFFICERS AND FAITHFUL
COMPANIONS WHO HAVE SUFFERED AND PERISHED
IN THE CAUSE OF SCIENCE AND
THE SERVICE OF THEIR COUNTRY.

THIS TABLET

IS ERECTED NEAR THE SPOT WHERE
THEY PASSED THEIR FIRST ARCTIC
WINTER, AND WHENCE THEY ISSUED
FORTH TO CONQUER DIFFICULTIES OR

TO DIE.

IT COMMEMORATES THE GRIEF OF THEIR
ADMIRING COUNTRYMEN AND FRIENDS,
AND THE ANGUISH, SUBDUED BY FAITH,
OF HER WHO HAS LOST, IN THE HEROIC
LEADER OF THE EXPEDITION, THE MOST
DEVOTED AND AFFECTIONATE OF
HUSBANDS.

---

*"AND SO HE BRINGETH THEM UNTO THE
HAVEN WHERE THEY WOULD BE."*

1855.

This stone has been intrusted to be affixed in its place by the Officers and Crew of the American Expedition, commanded by Lt. H. J. Hartstein, in search of Dr. Kane and his Companions.

This Tablet having been left at Disco by the American Expedition, which was unable to reach Beechey Island, in 1855, was put on board the Discovery Yacht Fox, and is now set up here by Captain M'Clintock, R.N., commanding the final expedition of search for ascertaining the fate of Sir John Franklin and his companions. 1858.

We are now ready to proceed upon our voyage from Beechey Island, and there is no ice in sight; but having worked almost unceasingly since our arrival up to the present hour, the men require a night's rest. Nearly forty tons of fuel have been embarked.

The total absence of ice in Barrow Strait is astonishing. No less so are the changes and chances of this singular navigation. Twelve days later than this in 1850, when I belonged to Her Majesty's ship 'Assistance,' with considerable difficulty we came within sight of Beechey Island: a cairn on its summit attracted notice; Captain Ommanney managed to land, and discovered the *first traces* of the missing expedition. Next day the United States schooner 'Rescue' arrived; the day after, Captain Penny joined us, and subsequently Captain Austin, Sir John Ross, and Captain Forsyth,—in all, ten vessels were assembled here. *This day* six years, when in command of the 'Intrepid,' we sailed from here for Melville Island in company with the 'Resolute.' Again I was here at this time in 1854,—still frozen up,—in the 'North Star,' and doubts were entertained of the possibility of *escape*.

To come down to a later period, it was this day fortnight only that I set out for the native village in Pond's Inlet, under the guidance of an old woman: the trip was interesting, but we failed to obtain the slightest clue to the "whereabouts" of the missing ships; moreover, our own little vessel had a most providential escape from being crushed against the cliffs; and this day week was spent in contending with a furious gale, during which the ship had nearly been driven to leeward and dashed to pieces by the sea-beaten pack. Yet these are only preliminaries,—we are only *now* about to commence the interesting part of our voyage. It is to be hoped the poor 'Fox' has many more lives to spare.

Monday night, 16th Aug.—Sailed from Beechey Island this morning, and in the evening landed at Cape Hotham. A small depôt of provisions and three boats were left there by former expeditions. Of the depôt all has been destroyed with the exception of two casks landed in 1850. The boats were sound, but several of their oars, which had been secured upright, were found broken down by bears—those inquisitive animals having a decided antipathy to anything stuck up—stuck-up things in general being, in this country, unnatural. Fragments of the depôt and the broken oars were tossed about in every direction. Numerous records were found; to the most recent a few lines were added, stating that we had removed the two whale-boats—one to be left at Port Leopold, the other to replace our own crushed by the ice.

17*th.*—Last night battling against a strong foul wind with *sea*, in rain and fog. To-day much loose ice is seen southward of Griffith's Island. The weather improved this afternoon, and we shot gallantly past Limestone Island, and are now steering down Peel Strait: all of us in a wild state of excitement—a mingling of anxious hopes and fears!

18*th.*—For 25 miles last evening we ran unobstructedly down Peel Strait, but then came in sight of unbroken ice extending across it from shore to shore! It was much decayed, and of one year's growth only; yet as the strait continues to contract for 60 miles further, and it appeared to me to afford so little hope of becoming navigable in the short remainder of the season, I immediately turned about for Bellot Strait, as affording a better prospect of a passage into the western sea discovered by Sir James Ross from Four River Point in 1849. Our disappointment at the interruption of our progress was as sudden as it was severe. We did not linger in hope of a change, but steered out again into the broad waters of Barrow Strait. However, should Bellot Strait prove hopeless, I intend to return hither to make one more effort before the close of the season.

We are now approaching Port Leopold, where it is necessary to stop for a few hours to examine the state of the steam launch, provisions, and stores, left there in 1849, as adverse circumstances may oblige me to fall back upon it as a point of support.

19*th.*—At anchor in Port Leopold; it is perfectly clear of ice; we arrived here in the night. How astonishingly bare the land looks; it is more barren than Beechey Island, whilst the rock contains far fewer fossils! On this day nine years ago the harbour and sea continued covered with ice, and the ships ('Enterprise' and 'Investigator') were unable to escape. At some period since then the ice has been pressed in upon the low shingle point; it has forced the launch up before it, and left her broadside on to the beach, with both bows stove in, and in want of considerable repairs, but the means are all at hand for executing them. We tried to haul her further up, but she was firmly imbedded and frozen into the ground. Many things appear to have been covered with the loose shingle, bags of coal and coke just appearing through it scarcely above high-water mark. Amongst the missing articles is the steam engine.

Although the flagstaff upon the summit of North East Cape is still standing, the one erected upon this point and almost the whole of the framing of the house lies prostrate. The provisions appeared to be sound, but were not generally examined. The whale-boat we removed from Cape Hotham was landed here, and a record of our proceedings added to the many which have accumulated here during the last ten years. Some coke and a few things useful to us and merely decaying here were taken on board, and by evening we were again speeding onward with augmented resources, and the confidence inspired by a secure depôt in our rear; buoyed up moreover by the joyful anticipation of soon reaching the goal of our long-deferred hopes.

20*th.*—Noon. Exactly off Fury Point. There is one large iceberg far off in the S.E.; no other ice in sight! I would have landed at Fury Beach to examine the remaining supplies there, but a snow shower prevented our distinguishing anything, and a strong tide carried us past before we were aware of it.

We *feel* that the crisis of our voyage is near

at hand. Does Bellot Strait really exist? if so, is it free from ice?

A depôt of provisions is being got ready to be landed, should it be practicable for us to push through and proceed to the southward.

21*st.*—On approaching Brentford Bay last evening, packed ice was seen streaming out of it, also much ice in the S.E. The northern point of entrance was landed upon by Sir John Ross in 1829, and named Possession Point; we rounded it closely, and could distinguish a few stones piled up upon a large rock near its highest part—this is his cairn. As we passed westward between the point and Browne's Island, through a channel a mile in width, a close pack was discovered a few miles ahead; and it being past ten o'clock, and almost dark, the ship was anchored in a convenient bay three or four miles within Possession Point. Here our depôt is to be landed, therefore we shall name this for the present *Depôt Bay*; a very narrow isthmus between its head and Hazard Inlet unites the low limestone peninsula, of which Possession Point is the extreme, to the mainland.

To-day an unsparing use of steam and canvas forced the ship eight miles further west; we were then about half-way through Bellot Strait! Its western capes are lofty bluffs, such as may be distinguished fifty miles distant in clear weather; between them there was a clear broad channel, but five or six miles of close heavy pack intervened—the sole obstacle to our progress. Of course this pack will speedily disperse;—it is no wonder that we should feel elated at such a glorious prospect, and content to bide our time in the security of Depôt Bay. A feeling of tranquillity—of earnest, hearty satisfaction—has come over us. There is no appearance amongst us of anything boastful; we have all experienced too keenly the vicissitudes of Arctic voyaging to admit of such a feeling.

At the turn of tide we perceived that we were being carried, together with the pack, back to the eastward; every moment our velocity was increased, and presently we were dismayed at seeing grounded ice near us, but were very quickly swept past it at the rate of nearly six miles an hour, though within 200 yards of the rocks, and of instant destruction! As soon as we possibly could, we got clear of the packed ice, and left it to be wildly hurled about by various whirlpools and rushes of the tide, until finally carried out into Brentford Bay. The ice-masses were large, and dashed violently against each other, and the rocks lay at some distance off the southern shore; we had a fortunate escape from such dangerous company. After anchoring again in Depôt Bay, a large stock of provisions and a record of our proceedings were landed, as there seems every probability of advancing into the western sea in a very few days.

The appearance of Bellot Strait is precisely that of a Greenland fiord; it is about 20 miles long and scarcely a mile wide in the narrowest part, and there, within a quarter of a mile of the north shore, the depth was ascertained to be 400 feet. Its granitic shores are bold and lofty, with a very respectable sprinkling of vegetation for lat. 72°. Some of the hill-ranges rise to about 1500 or 1600 feet above the sea.

The low land eastward of Depôt Bay is composed of limestone, destitute alike of fossils and vegetation. The granite commences upon the west shore of Depôt Bay, and is at once bold and rugged. Many seals have been seen; a young bear was shot, and Walker took a photograph of him as he lay upon our deck, the dogs creeping near to lick up the blood.

The great rapidity of the tides in Bellot

Strait fully accounts for the spaces of open water seen by Mr. Kennedy* when he travelled through, early in April. The strait runs very nearly east and west, but its eastern entrance is well masked by Long Island; when half-way through both seas are visible. As in Greenland, the night tides are much higher than the day tides; last night it was high water at about half-past eleven; as nearly as we can estimate, the tide runs through to the west, from two hours before high water until four hours after it; that is, the flood-tide comes from the west! Such is also the case in Hecla and Fury Strait; in both places the tide from the west is much the strongest. I am not sufficiently informed to discuss this subject, but infer the existence of a channel between Victoria and Prince of Wales' Land. The rise and fall is much less upon the western side of the Isthmus of Boothia than upon the east, and it likewise decreases, we know, in Barrow Strait, as we advance westward.

23*rd.*—Yesterday Bellot Strait was again examined, but the five miles of close pack occupied precisely the same position as if heaped together by contending tides; consider-

* Mr. Kennedy discovered this important passage when in command of the 'Prince Albert' in 1851.

able augmentations were moreover seen drifting in from the western sea. Finding nothing could be effected in Bellot Strait, we sought in vain for the more southern channel which should exist to form Levesque Island: we did, however, find a beautiful harbour, and are now securely anchored in its north-west arm; I have named it after the gentleman whose former island I have thus reluctantly converted into the northern extreme of the Boothian Peninsula, and consequently of the American continent. The south-western angle of Brentford Bay is still covered with unbroken ice.

This evening we all landed to explore our new ground. Young and Petersen shot some brent geese; Walker saw two deer, but he was botanising, and had no gun; others were seen by some of the men, and followed, but without success.

I enjoyed a delightfully refreshing ramble, a mile or two inland, through a gently ascending valley, then two miles along the narrow margin of a pretty little lake between mountains, beyond which lay a much larger one, four or five miles in diameter; this farther lake was only partially divested of its winter ice. Here the scenery was not only grand, but beautiful; there was enough of vegetation to tint the craggy hill-sides and to make the sheltered hollows absolutely green; deer-tracks and the footprints of wild-fowl were everywhere numerous along the water-side. I saw two decayed skulls of musk oxen, and circles of stones by the little lake, doubtless at some remote period the summer residence of wandering Esquimaux: hence I infer that fish abound in the lake, and that this valley is a favourite deer-pass.

But the contemplation of these objects, although agreeable, was not the object of my solitary ramble: I came on shore to cogitate undisturbed in a leisurely and philosophic manner. We hoped very soon to enter an unknown sea: discoveries were to be made, contingencies provided for, and plans prepared to meet them.

Yesterday Petersen shot an immense bearded seal; it sank, but floated up an hour afterwards. This animal measured 8 feet long, and weighed about 500 lbs. We prefer its flesh to that of the small seals, and its blubber will afford a valuable addition to our stock of lamp oil for the coming winter.

25*th.*—In Depôt Bay. We remained but twenty-four hours in Levesque Harbour; a change of wind led us to hope for a removal of the ice in Bellot Strait, therefore I determined to make another attempt.

When off the table-land, where the depth is not more than from 6 to 10 fathoms, and the tides run strongest, the ship hardly moved over the ground, although going 6½ knots through the water! Thus delayed, darkness overtook us, and we anchored at midnight in a small indentation of the north shore, christened by the men *Fox's Hole*, rather more than half-way through.

For several hours we had been coquetting with huge rampant ice-masses that wildly surged about in the tideway, or we dashed through boiling eddies, and sometimes almost grazed the tall cliffs; we were therefore naturally glad of a couple or three hours' rest, even in such a very unsafe position. At early dawn we again proceeded west, but for three miles only; the pack again stopped us, and we could perceive that the western sea was covered with ice; the east wind, which could alone remove it, now gave place to a hard-hearted westerly one.

All the strait to the eastward of us, and the eastern sea, as far as could be seen from the hill-tops, is perfectly free from ice, whereas in the direction we wish to proceed there is nothing but packed-ice, or water which cannot be reached. Bitterly disappointed we are, of course; yet there is reasonable ground for hope; grim winter will not ratify the obstinate proceedings of the western ice for nearly four weeks.

Last evening's *amusement* was most exciting, nor was it without its peculiar perils. With cunning and activity worthy of her name, our little craft warily avoided a tilting-match with the stout blue masses which whirled about, as if with wilful impetuosity, through the narrow channel; some of them were so large as to ground even in 6 or 7 fathoms water. Many were drawn into the eddies, and, acquiring considerable velocity in a contrary direction, suddenly broke bounds, charging out into the stream and entering into mighty conflict with their fellows. After such a frolic the masses would revolve peaceably or unite with the pack, and await quietly their certain dissolution; may the day of that wished-for dissolution be near at hand! Nothing but strong hope of success induced me to encounter such dangerous opposition. I not only hoped, but almost felt, that we deserved to succeed.

Two plans were now occupying my thoughts, both of them resulting from the conviction that we should probably be compelled to winter to the eastward of Bellot Strait: the most important of these plans is that of finding some series of valleys, chain of lakes, or continuous low land, practicable as an overland sledge-route to the western coast, along which we may transport depôts of provisions this autumn; for it is certain that the strong tides will prevent Bellot Strait being frozen over till winter is far advanced, and its surface will afford us no means of passing westward with our sledges.

The other plan, and that which we are now about to execute, is to land a small depôt of provisions 60 or 70 miles to the southward, and down Prince Regent's Inlet, in order to facilitate communication with the Esquimaux either this autumn or in early spring.

This precautionary step became so necessary in the event of the west coast presenting unusual difficulties, that I determined to carry it at once into execution. Quitting the "Fox's Hole," and resting for one night in Depôt Bay, we sailed thence on the 26th; a fine breeze carried us rapidly southward along the coast of Regent Inlet; there was but little obstruction; occasionally it was necessary to pass through a stream of loose ice; but we saw little of any kind, compared to the experiences of Sir John Ross in 1829.

About dusk (nine o'clock) much loose ice to the southward prevented our making any attempt at further progress; we therefore anchored off the coast—in Stillwell Bay, I think—about 45 miles from Depôt Bay. Here the depôt, consisting of 120 rations, was landed. I observe that it has only been on penetrating into Brentford Bay that we have found the primary rocks washed by the sea; the coast-line both north and south, as far as, and beyond our present position, is a low shore of pale limestone, destitute of fossils; we can however see granitic hill-ranges far in the interior.

On the 27th we commenced beating back to the northward, tacking between the land and the ice which lay about 15 miles off shore. Towards night the wind greatly increased, and the ship, under reefed sails, plunged violently into the short, swift, high seas; we also felt quite as uneasy and restless as the ship, in our great anxiety to get back and ascertain what changes were likely to be effected by the gale.

28*th.*—To-night the weather is more pleasant; the keen and contrary wind has given place to a gentle fair breeze, the swell has almost subsided, no ice has been seen to-day, and the night is dark and unusually mild. I can hardly fancy that the sea which gently rocks us is not the ocean, and the soft air the breath of our own temperate region! The delusion is charming.

30*th*.—Yesterday, after anchoring in Depôt Bay, I walked over to Possession Point, to visit Ross's cairn. I found a few stones piled up on two large boulders, and under each a halfpenny, one of which I pocketed. Upon the ground lay the fragments of a bottle which once contained the record, and near it a staff about 4 feet long. Having calculated upon finding the bottle sound, I was obliged to make an impromptu record-case of its long neck, into which I thrust my brief document, and consigned it to the safe custody of a small heap of stones, the staff being erected over it.

It was dark before I got on board again. The strait had been reconnoitred from the hills, and was reported to be perfectly clear of ice! This morning we made a fourth attempt to pass through; but Bellot Strait was by no means clear; the same obstruction existed which defeated our last attempt, and in precisely the same place. Returning eastward, we entered a narrow arm of the sea, nearly a couple of miles to the west of Depôt Bay, and anchored in a small creek, perfectly sheltered and land-locked, at the foot of a sugarloaf hill* The temperature is falling; last night it stood at 24°.

---

* Subsequently named Mount Walker.

CHAPTER XI.

Proceed westward in a boat — Cheerless state of the western sea — Struggles in Bellot Strait — Falcons, good Arctic fare — The resources of Boothia Felix — Future sledge travelling — Heavy gales — Hobson's party start — Winter quarters — Bellot's Strait — Advanced depôt established — Observatories — Intense cold — Autumn travellers — Narrow escape.

MOST anxious to know the real state of the ice in the western sea—upon which our hopes so entirely depend — I intend starting this evening by boat, as far through Bellot Strait as the ice will permit, then land and ascend the western coast-hills.

1*st Sept.*—My boat party consisted of four men and the Doctor, who came with me for the novelty of the cruise, bringing his camera to fasten upon anything picturesque. We landed near Half-way Island, and pitched our tent for the night. Early next morning I commenced the rather formidable undertaking of ascending the hills, for it is not possible to pass under the cliffs, and at last I gained the summit of the loftiest, overlooking Cape Bird at a distance of 3 or 4 miles, and affording a splendid view to

M'CLINTOCK IN HIS BOAT SAILING THROUGH BELLOT STRAIT

Drawn by Captain May

the westward, as well as glimpses between the hills of the blue eastern sea. Long and anxiously did I survey the western sea, ice, and lands, and could not but feel that in all probability we should not be permitted to pass beyond our present position.

To the northward Four River Point—Sir James Ross's farthest in 1849—was at once recognised; rather more than nine years ago I stood upon it with him, and gazed almost as anxiously in this direction! My present view confirmed the impression then received, of a wide channel leading southward. The outline of the western land is very distant; it is of considerable but uniform elevation, and slopes gradually down to the strait, which is between 30 and 40 miles wide. This western land appears to be limestone, and without offlying islands. Our side of the strait or sea, on the contrary, is primary rock, and fringed with islets and rocks; its southern extreme bears S.S.W., and is probably 30 miles distant.

Now for the ice. Although broken up, it lies against this shore in immense fields: there is but little water or room for ice-movement. Along the west shore I can distinguish long faint streaks of water. There is no appearance of disruption about Four River Point or in the contracted part of Peel Strait—we have nothing to hope for in that quarter; neither is there any evidence of current or pressure; the ice appears much decayed, but, as I am surveying it from a height of about 1600 feet, I may be deceived.

The strong contrast between the eastern and western seas and lands is very unfavourable to the latter.

Apart from the ice, I was fortunate, however, in discovering a long narrow lake, occupying a valley which lies between a small inlet near Cape Bird and Hazard Inlet—in fact, a sort of echo of Bellot Strait—and I look upon it as our sledge-route for the autumn, since it appears probable we shall winter in our present position.

This is a *wondrous rough* country to scramble over; one never ceases to wonder how such huge blocks of rock can have got into such strange positions. I noticed two masses in particular, each of them perched upon three small stones. The rock is gneiss; there is also much granite. Even upon the hill-tops pieces of limestone are occasionally met with.

My walk occupied eleven hours, and, although I everywhere saw traces of animals, the only living thing seen was a grey falcon. During my absence from the tent the men rambled all over the hills, but saw no game, our encampment was therefore shifted to a better position near the eastern termination of the table-land. This morning we explored the neighbouring valleys; saw three deer, and shot one, returning on board the 'Fox' in time for dinner.

Many deer had been seen not far from the ship, and Hobson had shot a bearded seal. I have organized another boat party; Young will start with it to-morrow morning to seek a sledge route from the southern angle of Brentford Bay to the western sea.

5th.—Young returned this morning; he reports the south-west angle of the bay not to run in so far as we expected, and to be environed by very high land, impracticable for sledges.

Our Esquimaux, Samuel, shot a fawn to-day.

Strong northerly winds have latterly prevailed; Bellot Strait is quite clear of ice; tomorrow morning, therefore, we shall make our *fifth* attempt to get the 'Fox' through.

6th.—Steamed through the clear waters of Bellot Strait this morning, and made fast to the ice across its western outlet at a distance of two miles from the shore, and close to a small islet which we have already dubbed *Pemmican Rock*, having landed upon it a large supply of that substantial traveller's fare, with other provisions for our future sledging-parties. This ice is in large stout fields, of more than one winter's growth, apparently immovable in consequence of the numerous islets and rocks which rise through and hold it fast. If the weather permits, we shall remain here for a few days and watch the effect of winds and tides upon it; that the ship will get any further seems improbable.

10*th*.—I have explored a small inlet near Cape Bird, which we have named *False Strait*, from its striking resemblance to the true one, and find it is only separated from the long lake by half-a-mile of low land; the lake we have ascertained to be about 12 miles long, and from it valleys extend eastward and southward, so that we are sure of a good sledge route,—an important matter, as the hills rise to 1600 feet above the sea.

Cape Bird is 500 feet high; from its summit we carefully observe the ice. This granite coast presents a jagged appearance; it is deeply indented and studded with islets. The ice in the western sea (or Peel Strait) is much more broken up than it was upon the 31st ultimo; there is no longer any fixed ice except within the grasp of the islets. Birds and animals have become very scarce; three seals have been shot, and a bear seen. To-morrow we shall return to our harbour, and endeavour to procure a few more reindeer before they migrate southward.

12*th*.—Yesterday we anchored within the entrance of our creek, being a more convenient position than up at its head. We are already in our wintering position, and, being without occupation, one day seems most remarkably like another! Although the fondly cherished hope of pushing farther in our ship can no longer be entertained, yet as long as the season continues navigable, it is our duty to be in readiness to avail ourselves of any opportunity, however improbable, of being able to do so.

Once firmly frozen in, our autumn travelling will commence, and afford welcome occupation. Almost all on board have guns; ammunition is supplied, and a sailor with a musket is a very contented and zealous sportsman, if not always a successful one; it is a powerful incentive to exercise. To-day the ramblers saw only two hares, an ermine, and an owl. Some peregrine falcons have lately been shot; Petersen declares they are "*the best beef in the country, and the young birds tender and white as chicken!*"

A few days ago a large cask of biscuit was opened, and a living mouse discovered therein!

it was small, but mature in years. The cask, a strong watertight one, was packed on shore at Aberdeen in June, 1857, and remained ever afterwards unopened; there was no hole by which the mouse could have got in or out, besides it is the only one ever seen on board. Ships' biscuit is certainly *dry feeding*, but who dares assert, after the experience of our mouse, that it is not wonderfully nutritious?

15*th*.—Two nights ago a comet was observed just beneath the constellation of the Great Bear; a series of measurements were commenced for determining its path. Yesterday I walked through the most promising valleys for eight hours, but did not see a living creature; yet there is a very fair show of vegetation, much more than at Melville Island, where the game is abundant. To the east there is not a speck of ice, excepting only a huge iceberg, probably the same we saw off Fury Point, a very unusual visitor from Baffin's Bay, whence it must have been driven by those long-continued east winds (of painful memory) in June and July.

Hobson and two men encamped out for three days in order to scour the country; they have only seen one hare and one lemming! Walker geologizes; amongst other things he finds much iron pyrites. The dredge has been used, but with very little success. The thermometer ranges between 20° and 30°. Fresh water pools are frozen over, sea-ice forms in every sheltered angle of the creeks. There is no snow upon the land, and this is one cause of the difficulty of finding game.

I have determined upon naming this beautiful little anchorage *Port Kennedy*, after my predecessor, the discoverer of Bellot Strait, of which it is decidedly *the* port. This is not a compliment to him, but an agreeable duty to me, and nowhere could Mr. Kennedy's name be more appropriately affixed than in close proximity with his interesting discovery. And now having made this acknowledgement, I may venture to confer our little vessel's name upon the islets which protect its entrance.

The island upon which Mr. Kennedy and Lieutenant Bellot encamped was Long Island, about three miles further to the south-east.

17*th.*—Of late we have been preparing provisions and equipments for our travelling parties. My scheme of sledge search comprehends three separate routes and parties of four men; to each party a dog sledge and driver will be attached; Hobson, Young, and I will lead them.

My journey will be to the Great Fish River, examining the shores of King William's Land in going and returning; Petersen will be with me.

Hobson will explore the western coast of Boothia as far as the magnetic pole, this autumn, I hope, and from Gateshead Island westward next spring.

Young will trace the shore of Prince of Wales' Land from Lieutenant Browne's farthest, to the southwestward to Osborn's farthest, if possible, and also examine between Four River Point and Cape Bird.

Our probable absence will be sixty or seventy days, commencing from about the 20th March.

In this way I trust we shall complete the Franklin search and the geographical discovery of Arctic America, both left unfinished by the former expeditions; and in so doing we can hardly fail to obtain some trace, some relic, or, it may be, important records of those whose mysterious fate it is the great object of our labours to discover. But previous to setting forth upon these important journeys, I must communicate with the Boothians, if possible, either upon the west or east coast, in November or February. Sir John Ross's 'Narrative' informs us that they sometimes winter as far

DOG SLEDGE OR SCOUT PARTY

north upon the east coast as the Agnew River; and we know that upon the west, at the magnetic pole, their abandoned snow huts were occupied in June by Sir James Ross.

19*th.* — Yesterday we steamed once more through Bellot Strait, and took up our former position at the ice-edge, off its western entrance; the ice, hemmed in by islets, has not moved.

From the summit of Cape Bird I had a very extensive view this morning; there is now much water in the offing, only separated from us by the belt of islet-girt ice *scarcely four miles in width!* My conviction is that a strong east wind would remove this remaining barrier; it is not yet too late. The water runs parallel to this coast, and is four or five miles broad; beyond it there is ice, but it appears to be all broken up.

Yesterday Young went upon a dog-sledge to the nearest south-western island, distant 7 or 8 miles. He reports the intervening ice cracked and weak in some places, but practicable for loaded sledges; the far side of the island is washed by a clear sea, and a bear which he shot plunged into it, and, drifting away, was lost. Young is in favour of carrying out the depôt provisions to or beyond this island by boat; but as the temperature fell to 18° last night, and new ice forms whenever it is calm, I prefer the safer, although more laborious mode of sledging; accordingly to-day our dogs carried out two sledge-loads of the provisions intended for the use of our parties hereafter.

22nd.—All the provisions have now been carried out to the nearest island, which I shall temporarily name *Separation*,* as there our spring parties will divide; and a portion intended for Hobson's party and my own has been carried on to the next island 7 or 8 miles further. Our travelling boat and a small reserve depôt have been placed upon Pemmican Rock, so already something has been done. Animal life is very scarce; a few seals, an occasional gull, and three brown falcons, are the only creatures we have seen for several days past. Last evening at eight o'clock a very vivid flash of lightning was observed; its appearance in these latitudes is very rare; once only have I seen it before—in September, 1850.

25th.—Saturday night. Furious gales from N. and S.W., but our barrier of coast-ice remains undiminished. This morning Hobson set off upon a journey of fourteen or fifteen days' duration, with seven men and fourteen dogs;

* Subsequently named after my excellent friend A. Arcedeckne, Esq., commodore of the Royal London Yacht Club.

he is to advance the depôts along shore to the south, and if successful will reach latitude 71°.

The temperature is mild (+17), but it is snowy and disagreeable weather; there is already enough snow upon the old ice to make walking laborious, and the land has also assumed its wintry complexion.

28*th*.—The ship was kept available for prosecuting her voyage up to the *latest hour*; it was only yesterday that we left the western ice, and in consequence of the vast accumulation of young ice in Bellot Strait we had considerable difficulty in reaching the *entrance* of Port Kennedy: all within was so firmly frozen over that after three hours' steaming and working we only penetrated 100 yards; however, we are in an excellent position, although our wintering place will be farther out by a quarter of a mile than I intended.

To-day we are unbending sails and laying up the engines—uncertainty no longer exists—here we are compelled to remain; and if we have not been as successful in our voyaging as a month ago we had good reason to expect, we may still hope that Fortune will smile upon our more humble, yet more arduous, pedestrian explorations—"Hope on, hope ever." In the mean time the sudden transition, from mental and physical wear and tear, to the security and quiet of winter quarters, is an immense relief.

2nd Oct.—Mr. Petersen has shot two very fine bucks; one is a magnificent fellow, weighing 354 lbs. (minus the paunch). Several deer have been seen; they come from the N. along the slopes of the eastern hills. An ermine came on board a few nights ago and kept the dogs in a violent state of excitement, being much too wary to come out from under the boat to be caught by them; at length one of the men secured it. This beautiful little animal does not appear to be full grown; its extreme length is 13 inches. Two others came off to the ship, and to our great amusement eluded the men who gave chase, by darting into the soft snow—which is now a foot deep—and reappearing several yards off.

The weather is too mild to satisfy us; we wish for severe frost to seal us up securely, and make the ice strong enough to bear the sledge-loads of provisions, &c., which are to be landed for the purpose of making more room in the ship.

6th.—A herd of a dozen reindeer crossed the harbour to-day. Last night Hobson and his companions returned, all well. They were stopped by the sea washing against the cliffs in latitude 71½°, and to that point they have advanced the depôts. Although the weather has been stormy here, they have been able to travel every day. They found the coast still fringed with islets, and deeply indented; upon every point, moss-grown circles of stones indicated the abodes of Esquimaux in times long since gone by.

One night they muzzled a dog, as she was in the habit of gnawing her harness: in this defenceless state, unable even to bark and arouse the men, her *amiable* sisterhood attacked her so fiercely that she died next day!

In honour of so important and successful a commencement of our travelling, as that accomplished by Hobson, we had a feast of good venison, plum pudding, and grog. It is quite evident that no more travelling can be accomplished until the ice forms a pathway alongshore; in this, as in some other respects, we anxiously await the advance of the season. The weather is mild; Bellot Strait is almost covered with ice, which drifts freely with every tide. Reindeer are seen almost daily; they too are awaiting the freezing over of the sea to continue their southern travels. Our harbour-ice is weak and covered a foot deep with a sludgy compound of snow and water.

8*th.*—Yesterday an ermine was caught in a

trap; hitherto these most active little skirmishers have successfully robbed our fox-traps of their baits as fast as they could be renewed. To-day Petersen shot another reindeer; it weighs 130 lbs.; many others were seen, also a wolf. Sometimes a few ptarmigan are met with, but hares very rarely.

12*th.*—Fine weather generally prevails. We have landed about 100 casks, all our boats, and much lumber, so we shall have abundance of room on board. I enjoyed a long and exhilarating ramble upon snow-shoes to-day; without them I could not have gone over half the distance—the snow lies so deep and soft—but I only saw one reindeer.

14*th.*—One of our magnetic observatories has been built; it stands upon the ice, 210 yards S. (magnetic) from the ship, and is built of ice sawed into blocks—there not being any suitable snow; it is just large enough to hold the declinometer for hourly observations, to be noted throughout the winter. The housings have been put over the ship already, as Hobson will leave us again in a few days to advance his depôt and my own to the vicinity of the magnetic pole if possible. I would also send Young upon a similar duty, but the western sea cannot have frozen over yet.

INTERIOR OF THE OBSERVATORY

Drawn by Captain May.

19*th*.—All the 17th a N.W. gale blew with fearful violence; yesterday it abated, but not sufficiently to allow our party to start. This morning Hobson got away with his nine men and ten dogs; his absence may be from eighteen to twenty days. Autumn travelling is most disagreeable; there is so much wind and snow, the latter being soft, deep, and often wet; the sun is almost always obscured by mist, and is powerless for warmth or drying purposes, and the temperature is very variable. Moreover there are now only eight hours of misty daylight. To-day the morning was fine, and temperature + 8°. Having completed the preliminary observations of the times of horizontal and vertical vibrations, also of the magnetic intensity, I set up to-day the declinometer, and commenced the hourly series of observations on the diurnal variation. I trust it may continue unbroken until we all set out upon our spring travels in March. A hare has been shot, but no other animals seen.

29*th*.—It generally blows a gale of wind here; the only advantage in return for so much discomfort is that the snow is the more quickly packed hard. As we have only three working men and an Esquimaux left on board for ship's duties, I was assisted a few days ago by the

Doctor, the Engineer, and the Interpreter in building another observatory, intended for certain monthly magnetic observations. This edifice is constructed of snow. Whenever we have a calm night we can hear the crushing sound of the drift-ice in Bellot Strait, which continues open to within 500 yards of the Fox Islands, and emits dark chilling clouds of hateful, pestilent, abominable mist.

The last two days have been very fine and calm: the men visited their fox, and ermine, traps, which are secreted amongst the rocks in a most mysterious manner—one ermine only has been taken. Seven or eight reindeer and some ptarmigan were seen; two of the latter and a hare were shot. We have commenced brewing sugar beer.

2nd Nov.—Very dull times. No amount of ingenuity could make a diary worth the paper it is written on. An occasional raven flies past, a couple more ptarmigan have been shot: another N.W. gale is blowing, with temperature down to – 12°.

6th.—Saturday night. The N.W. gale blew without intermission for seventy hours, the temperature being about – 15°: we hoped that our absent shipmates might be housed safely in snow huts. This afternoon all doubts respecting them were dispelled by their arrival in good health, but they evidently have suffered from cold and exposure during their absence of nineteen days. For the first six days they journeyed outward successfully; on that night they encamped upon the ice; it was at spring-tide, a N.E. gale sprang up and blowing off shore, detached the ice and drifted them off! The sea froze over on the cessation of the gale, and two days afterwards they fortunately regained the land near the position from which they were blown off; they have indeed experienced much unusual danger and suffering from cold.

As soon as they discovered that the ice was drifting off shore with them, they packed their sledges, harnessed the dogs, and passed the night in anxious watching for some chance to escape. When the ice got a little distance off shore, it broke up under the influence of the wind and sea, until the piece they were upon was scarce 20 yards in diameter: this drifted across the mouth of a wide inlet* until brought up against the opposite shore. The gale was quickly followed by an intense frost, which in a

* Named after Lord Wrottesley, in remembrance of the support given by him to the expedition, his advocacy of it in the House of Lords, and of the facilities granted me by the Royal Society—of which he was President—for the pursuit of scientific observations.

single night formed ice sufficiently strong to bear them in safety to the land, although it bent fearfully beneath their weight.

The depôts were eventually established in latitude 71°; beyond this Lieutenant Hobson did not attempt to advance, not only because their remaining provisions would not have warranted a longer absence, but because the open sea was seen to beat against the next headland. They have lived in tents only, and have not experienced the heavy gales so frequent here, and which are probably due mainly to our position in Bellot Strait, which performs the part of a funnel for both winds and tides between the two seas.

That the western sea should still remain open argues a vast space southward for the escape of the ice, and prevents our western party from carrying across their depôt: the attempt to do so would be extremely hazardous. We must only be stirring earlier in the spring. I am truly thankful for the safe return of our travellers,—all this toil and exposure of ten persons and ten dogs has only advanced the depôts 30 miles further—*i. e.* from 60 to 90 miles distant from the ship.

Hardly a particle of snow remains upon the harbour-ice, the recent gales having swept it away; and the porch of my snow-hut has been fretted away to a mere cobweb by the attrition of the snowdrift: the Doctor and I rebuilt it to-day. Three reindeer and a wolf have been seen.

CHAPTER XII.

Death of our engineer — Scarcity of game — The cold unusually trying — Jolly, under adverse circumstances — Petersen's information — Return of the sun of 1859 — Early spring sledge parties — Unusual severity of the winter — Severe hardships of early sledging — The western shores of Boothia — Meet the Esquimaux — Intelligence of Franklin's ships — Return to the 'Fox' — Allen Young returns.

Nov. 7th.—Sunday evening. BRIEF as is the interval since my last entry, yet how awful and, to one of our small company, how fatal it has been! Yesterday Mr. Brand was out shooting as usual, and in robust health; in the evening Hobson sat with him for a little time. Mr. Brand turned the conversation upon our position and employments last year; he called to remembrance poor Robert Scott, then in sound health, and the fact of his having carried our "Guy Fawkes round the ship on the preceding day twelvemonth, and added mournfully, "Poor fellow! no one knows whose turn it may be to go next." He finished his evening pipe, and shut his cabin door shortly after nine o'clock. This morning, at seven o'clock, his servant found him lying upon the deck, a corpse, having been several hours dead. Apoplexy appears to have been the cause. He was a steady, serious man, under forty years of age, and leaves a widow and three or four children; what their circumstances are I am not aware.

10*th.*—This morning the remains of Mr. Brand, inclosed in a neat coffin, were buried in a grave on shore. A suitable headboard and inscription will be placed over it. From all that I have gathered, it appears that his mind had been somewhat gloomy for the last few days, dwelling much upon poor Scott's sudden death. Whether he really saw three reindeer on Saturday, watched their movements, and fired his Minié rifle at them when 700 yards distant, or whether it was the creation of a disordered brain, none can tell. On his first return on board he said he had seen deer *tracks* only.

We are now without either engineer or engine-driver: we have only two stokers, and they know nothing about the machinery. Our numbers are reduced to twenty-four, including our interpreter and two Greenland Esquimaux.

15*th.*—We have enjoyed ten days of moderate winds and calms, but the temperature has fallen as low as $-31°$. This causes frost-cracks in the ice *across* the harbour; they will freeze over, and others will form, and gape, and freeze at intervals, so that by next spring we shall probably be moved several inches, perhaps feet, off shore.

Mists have obscured the sun of late, and now it does not rise at all. We are indifferent: its departure has become to us a matter of course. The usual winter covering of snow has been spread upon deck rather more than a foot thick. Its utility in preventing the escape of heat became at once strikingly apparent. Nothing has been seen but a few ptarmigan and one reindeer, which trotted off towards the ship. Our bullets missed him, and the dogs unfortunately caught sight and chased him away. I do not think any dogs could overtake a reindeer in this rough country; the rocks would speedily lame them, and the snow, in many places, is quite deep enough to fatigue them greatly, whereas it offers but slight impediment to the deer, furnished as he is with long legs and spreading hoofs.

29*th.*—Animals have become very scarce. A few ptarmigan and willow-grouse have been seen, and three shot. Two days ago I saw two reindeer. The eastern sea is frozen over, and our old acquaintance the iceberg in Prince

Regent's Inlet is still visible on a clear day. We brew sugar-beer, and we set nets for seals, but catch none. The nets have been made and set in favourable positions under the ice by the Greenlanders, so we suppose the seals also have migrated elsewhere; if so, the Esquimaux could not winter here. We have no regular school this winter, but five of the men study navigation every evening under the guidance of Young. Hobson and I are doing all we can to make the ship dry, warm, and comfortable: our large snow porches over the hatchways are a great improvement.

5th Dec.—Cold, windy weather, with chilling mists from the open water in Bellot Strait. We can seldom leave the shelter of the ship for a walk on shore, and, when we do, rarely see even a ptarmigan.

12th.—Very cold weather; thermometer down to $-41°$, and the breeze comes to us loaded with mist from the open water, causing the air to feel colder than it otherwise would. Bellot Strait has become a nuisance, not only from this cause, but from the strong winds—purely local—which seldom cease to blow through it.

The seal nets have produced nothing; and as there are no seals, we no longer wonder at not seeing bears. Three foxes have been trapped and a hare seen. Our canine force numbers twenty-four serviceable dogs and six puppies; but these, I fear, will not be strong enough for sledging by March. The monotony of our lives is vastly increased by want of occupation, and confinement, by severe gales, to the ship for five days out of every seven. The general health is good, but there is a natural craving for fresh meat and fresh vegetables—in great measure, perhaps, because they cannot be obtained; but a well-filled letter-bag would be more welcome than anything I know of.

26*th.*—Upon four days only during the last fourteen has the weather permitted us to walk. I allude to the wind as the obstacle to our exercise; for temperature, when the air is still, is no bar to any reasonable amount of it. Three or four coveys of ptarmigan have been seen, and of these I shot one brace. The cold increases: thermometer has fallen to $-47\frac{1}{2}°$, although blowing a moderate gale at the time, and the atmosphere dense with mist.

Our Christmas has been spent with a degree of loyalty to the good old English custom at once spirited and refreshing. All the good things which could possibly be collected together appeared upon the snow-white deal tables of the men, as the officers and myself walked (by invitation) round the lower deck. Venison, beer, and a fresh stock of clay pipes, appeared to be the most prized luxuries; but the variety and abundance of the eatables, tastefully laid out, was such as might well support the delusion which all seemed desirous of imposing upon themselves — that they were in a land of plenty—in fact, *all but* at home! We contributed a large cheese and some preserves, and candles superseded the ordinary smoky lamps. With so many comforts, and the existence of so much genuine good feeling, their evening was a joyous one, enlivened also by songs and music.

Whilst all was order and merriment within the ship, the scene without was widely different. A fierce north-wester howled loudly through the rigging, the snowdrift rustled swiftly past, no star appeared through the oppressive gloom, and the thermometer varied between 76° and 80° *below the freezing point.* At one time it was impossible to visit the magnetic observatory, although only 210 yards distant, and with a rope stretched along, breast high, upon poles the whole way. The officers discharged this duty for the quartermasters of the watches during the day and night.

1*st Jan.* 1859.—This being *Saturday night* as well as *New Year's Day*, "Sweethearts and Wives" were remembered with even more than the ordinary feeling. New year's eve was celebrated with all the joyfulness which ardent hope can inspire: and we *have* reasonable ground for *strong hope*. At midnight the expiration of the old year and commencement of the new one was announced to me by *the band*—flutes, accordion, and gong—striking up at my door. Some songs were sung, and the performance concluded with "God save the Queen:" the few who could find space in our mess-room sang the chorus; but this by no means satisfied all the others who were without and unable to show themselves to the officers, so they echoed the chorus, and the effect was very pleasing. Our new year's day has been commemorated with all the substantials of Christmas fare, but without so much display,—less tailoring in pastry, not quite so much clipping of dough into roses, and anchors, and nondescript animals, &c. &c. The past week has been cold and stormy; it now blows strong, and the temperature is −44°.

On the 29th a few fresh tracks of animals

and a ptarmigan were seen: yesterday I saw three ptarmigan. December proved to be an unusually cold month, its mean temperature being $-33°$; and it was rendered more than ordinarily dark and gloomy by continual mists from Bellot Strait. This open water adds seriously to the drawbacks of a spot already sufficiently cheerless, gameless, and "wind-loved."

9th.—Another week of uniform temperature of $-40°$, and confinement to the ship by strong winds; the atmosphere is loaded with enveloping mists which impart a raw and surprisingly keen edge to the chilling blasts, blasts that no human nose can endure without blanching, be its proportions what they may. It is wonderful how the dogs stand it, and without apparent inconvenience, unless their fur happen to be thin. They lie upon the snow under the lee of the ship, with no other protection from the weather.

To-day, the winds being light and temperature *up to* $-30°$, we enjoyed walks on shore, although the mist continued so dense as to limit our view to a couple of hundred yards.

I learn from Petersen that the natives of Smith's Sound are well acquainted with the continuation of its shores considerably beyond the farthest point reached by Kane's exploring parties, but unfortunately no one thought of getting them to delineate their local knowledge upon paper. They spoke much of a large island near the west coast called "Umingmak" (musk ox) Island, where there was much open water, abounding with walrus, and where some of their people formerly lived.*

Esquimaux exist upon the east coast of Greenland as far north as lat. 76°; how much farther north is not known. They are separated from the South Greenlanders by hundreds of miles of icebound coasts and impassable glaciers.

Many centuries ago a milder climate *may* and probably *did* exist, and a corresponding modification of glacier and a sea less ice-encumbered might have rendered the migration of these poor people from the south to their present isolated abodes practicable; but to me it appears much more easy to suppose that they migrated eastward from the northern outlet of Smith's Sound.

21*st.*—More pleasant weather since my last entry; and although last night the temperature fell to −47°, yet it has generally been mild; once it rose to −14°, but amply made amends by falling to −38° within twelve hours. We have

* Petersen conversed with two men who had themselves been up to Umingmak Island.

WALRUSES—A FAMILY PARTY

From a Sketch by Captain Allen Young

enjoyed much of the moon's presence for the last ten days, but now she is waning and hastening away to the south. Daylight increases in strength and duration, consequently we walk more, and see more, and the winter's gloom gives place to activity and cheerfulness. Several ptarmigan, three or four hares, a snowy owl, and a bear-track, have at various times been seen. Young has shot four ptarmigan, and I have shot a couple more and a hare, and the men have trapped two foxes.

On board the ship the preparations for travelling take precedence of all other occupations.

26*th*.—Part of the sun's disc loomed above the horizon to-day, somewhat swollen and disfigured by the misty atmosphere, but looking benevolent withal. I happened to be diligently traversing the rocky hill-sides in the hope of finding some solitary hare dozing in fancied security, when the sun thus appeared in view, and halted to feast my eyes upon the glorious sight, and scan the features of our returning friend. Hope and promise mingled in his bright beams. Again I moved upward, and with more elastic step; for now the sun of 1859 was shining upon all nature around me.

2*nd February*.—A lovely, calm, bright day, and beautifully clear, except over the water-space in Bellot Strait, where rests a densely black mist, very strongly resembling the West Indian rain-squall as it looms upon the distant horizon. The increasing sunlight is cheering, but void of heat, and the mercury is often frozen. A few more ptarmigan have been shot.

Our remaining serviceable dogs, twenty-two in number, have been divided with great care into three teams of seven each; the odd dog is added to my team, as my journey is expected to be the longest. The different sledge-parties will now feed up their dogs without limit, so that the utmost degree of work may be got out of them hereafter.

January has been slightly colder than December, mean temperature being − 33½°, but there has been rather less wind.

8*th.*—All will be ready for the departure of Young and myself upon our respective journeys upon the morning of the 14th.

Mr. Petersen and Alexander Thompson accompany me, with two dog-sledges, and fifteen dogs, dragging twenty-four days' provisions. My object is to communicate with the Boothians in the vicinity of the magnetic pole. Young takes his party of four men and his dog-sledge; he will carry forward provisions for his spring exploration of the shores of Prince of Wales' Land, between the extreme points reached by Lieutenants Osborn and Browne in 1851.

On the 3rd I walked for seven and a half hours, and saw two reindeer, but could not approach within shot. Young examined the water-space in the strait, and finds it washes both shores, but extends east and west only about one mile. The Doctor has seen a seal and a dovekie sporting in it.

For the last four days strong winds and intense cold have prevented us from rambling over the hills, besides which the minor preparations for travelling have given us more occupation on board.

James Pitcher has got a slight touch of scurvy; his gums are inflamed; and now it comes out that he dislikes preserved meats, and has not eaten any since he has been in the ship! He has lived upon salt meat and preserved vegetables, except for the very short periods in summer when birds could be obtained. He is rather a "used-up" old fellow, too much so for our severe sledge-work, therefore is one of the few who will remain to take care of the ship. That he should have retained his health for seventeen months, under the circumstances, speaks well for the wholesomeness and quality of our provisions, and the ventilation and cleanliness of the ship.

10*th*.—Extremely cold, with dense mists from the open water. Yesterday eight ptarmigan and a sooty fox were seen. We have consumed the last of our venison; it supplied us for three days. We are drinking out a cask of sugar-beer, which is a very mild but agreeable beverage; we make it on board.

Sunday night, 13*th*.—To-morrow morning, if fine, Young and I set off upon our travels. He has advanced a portion of his sledge-load to the west side of the water in Bellot Strait, having been obliged to carry it overland for about a mile in order to get there. I have explored the route to the long lake, and find we can reach it without crossing elevated or uncovered land. I saw two reindeer, and Young saw about twenty ptarmigan.

The mean temperature of February up to this date is $-33 \cdot 2°$, being an exact continuation of January. I confess to some anxiety upon this point, as hitherto the winter has been unusually severe, and the journeys to be performed will occupy more than twenty days. Besides, we shall be earlier in motion than any of the previous travellers, unless we are to make an exception in favour of Mr. Kennedy's trip of 30 miles from Batty Bay to Fury Beach, between the 5th and 10th January, during which time the lowest temperature registered was only $-25°$. Should either Young or myself remain absent beyond the period for which we carry provisions, Hobson is to send a party in search of us. A sooty fox has been captured lately.

15*th.*—A strong N.W. wind, with a temperature of $-40°$, confines us on board. One cannot face these winds, therefore it is fortunate that we did not start, the ship being much more comfortable than a snow-hut.

* * * * *

20*th March.*—Already I have been a week on board, and so difficult is it to settle down to anything like sedentary occupation, after a period of continued vigorous action, that even now I can scarcely sit still to scribble a brief outline of my trip to Cape Victoria.

On the morning of the 17th February the weather moderated sufficiently for us to set out; the temperature throughout the day varied between $-31°$ and $-42\frac{1}{2}°$. Leaving Young's party to pass on through the strait, I proceeded by way of the Long Lake, which I found to be 10½ geographical miles in length, with an average width of half a mile.

We built our snow-hut upon the west coast,

near Pemmican Rock, after a march of 19 or 20 geographical miles. We always speak of *geographical* miles with reference to our marches; six geographical are equal to seven English miles.

On the following day the old N.W. wind sprang up with renewed vigour, and the thermometer fell to $-48°$; the cold was therefore intense.

On the third day most of our dogs went lame in consequence of sore feet; the intense cold seems to be the principal, if not the only cause, having hardened the surface-snow beyond what their feet can endure. I was obliged to throw off a part of the provisions; still we could not make more than 15 or 18 miles daily. We of course walked, so that the dogs had only the remaining provisions and clothing to drag, yet several of them repeatedly fell down in fits.

For several days this severe weather continued, the mercury of my artificial horizon remaining frozen (its freezing-point is $-39°$); and our rum, at first thick like treacle, required thawing latterly, when the more fluid and stronger part had been used. We travelled each day until dusk, and then were occupied for a couple of hours in building our snow-hut. The four walls were run up until 5½ feet high, inclining inwards as much as possible; over these our tent was laid to form a roof; we could not afford the time necessary to construct a dome of snow.

Our equipment consisted of a very small brown-holland tent, macintosh floor-cloth, and felt robes; besides this, each man had a bag of double blanketing, and a pair of fur boots, to sleep in. We wore mocassins over the pieces of blanket in which our feet were wrapped up, and, with the exception of a change of this foot-gear, carried no spare clothes. The daily routine was as follows:—I led the way; Petersen and Thompson followed, conducting their sledges; and in this manner we trudged on for eight or ten hours without halting, except when necessary to disentangle the dog-harness. When we halted for the night, Thompson and I usually sawed out the blocks of compact snow and carried them to Petersen, who acted as the master-mason in building the snow-hut: the hour and a half or two hours usually employed in erecting the edifice was the most disagreeable part of the day's labour, for, in addition to being already well tired and desiring repose, we became thoroughly chilled whilst standing about. When the hut was finished, the dogs were fed, and here the great difficulty was to insure the weaker ones their full share in the scramble for supper; then commenced the operation of unpacking the sledge, and carrying into our hut everything necessary for ourselves, such as provision and sleeping gear, as well as all boots, fur mittens, and even the sledge dog-harness, to prevent the dogs from eating them during our sleeping hours. The door was now blocked up with snow, the cooking-lamp lighted, foot-gear changed, diary written up, watches wound, sleeping bags wriggled into, pipes lighted, and the merits of the various dogs discussed, until supper was ready; the supper swallowed, the upper robe or coverlet was pulled over, and then to sleep.

Next morning came breakfast, a struggle to get into frozen mocassins, after which the sledges were packed, and another day's march commenced.

In these little huts we usually slept warm enough, although latterly, when our blankets and clothes became loaded with ice, we felt the cold severely. When our low doorway was carefully blocked up with snow, and the cooking-lamp alight, the temperature quickly rose so that the walls became glazed, and our bedding thawed; but the cooking over, or the doorway partially opened, it as quickly fell again, so that it was impossible to sleep, or even to hold one's pannikin of tea, without putting our mitts on, so intense was the cold!

On the 21st I visited our main depôt laid out last October; it was safe, but unfortunately had been carried far into Wrottesley Inlet, and only 40 miles south of Bellot Strait.

On the 22nd an easterly gale prevented our marching, but we had the good fortune to shoot a bear, so consoled ourselves with fresh steaks, and the dogs with an ample feed of *unfrozen* flesh—a treat they had not enjoyed for many months.

We coasted along a granitic land, deeply indented and fringed with islands, and found it to be the general characteristic of the Boothian shore from Bellot Strait, until we had accomplished half the distance to the magnetic pole; limestone then appeared, and the remainder of our journey was performed along a low, straight shore, which afforded us much greater facility for sledging.

Throughout the whole distance we found a mixture of heavy old ice and light ice of last autumn, in many places squeezed up into pack; but as we advanced southward aged floes were less frequently seen.

On the 1st of March we halted to encamp at about the position of the magnetic pole—for no cairn remains to mark the spot. I had almost concluded that my journey would prove to be a work of labour in vain, because hitherto no traces of Esquimaux had been met with, and, in consequence of the reduced state of our provisions and the wretched condition of the poor dogs—six out of the fifteen being quite useless—I could only advance one more march.

But we had done nothing more than look *ahead*; when we halted, and turned round, great indeed was my surprise and joy to see four men walking after us. Petersen and I immediately buckled on our revolvers and advanced to meet them. The natives halted, made fast their dogs, laid down their spears, and received us without any evidence of surprise. They told us they had been out upon a seal hunt on the ice, and were returning home: we proposed to join them, and all were soon in motion again; but another hour brought sunset, and we learned that their snow village of eight huts was still a long way off, so we hired them, at the rate of a needle for each Esquimaux, to build us a hut, which they completed in an hour; it was 8 feet in diameter, 5½ feet high, and in it we all passed the night.

Perhaps the records of architecture do not furnish another instance of a dwelling-house so cheaply constructed!

We gave them to understand that we were anxious to barter with them, and very cautiously approached the real object of our visit. A naval button upon one of their dresses afforded the opportunity; it came, they said, from some white people who were starved upon an island where there are salmon (that is, in a river); and that the iron of which their knives were made came from the same place. One of these men said he had been to the island to obtain wood and iron, but none of them had seen the white men. Another man had been to "Ei-wil-lik" (Repulse Bay), and counted on his fingers seven individuals of Rae's party whom he remembered having seen.

These Esquimaux had nothing to eat, and no other clothing than their ordinary double dresses of fur; they would not eat our biscuit or salt pork, but took a small quantity of bear's blubber and some water. They slept in a sitting posture, with their heads leaning forward on their breasts. Next morning we travelled about 10 miles further, by which time we were close to Cape Victoria; beyond this I would not go, much as they wished to lead us on; we therefore landed, and they built us a commodious snow hut in half an hour; this done, we displayed to them our articles for barter—knives, files, needles, scissors, beads, &c.—expressed our desire to trade with them, and promised to purchase everything which belonged to the starved white men, if they would come to us on the morrow. Notwithstanding that the weather was now stormy and bitterly cold, two of the natives stripped off their outer coats of reindeer skin and bartered them for a knife each.

Despite the gale which howled outside, we spent a comfortable night in our roomy hut.

Next morning the entire village population arrived, amounting to about forty-five souls, from aged people to infants in arms, and bartering commenced very briskly. First of all we purchased all the relics of the lost expedition, consisting of six silver spoons and forks, a silver medal, the property of Mr. A. M'Donald, assistant surgeon, part of a gold chain, several buttons, and knives made of the iron and wood of the wreck, also bows and arrows constructed of materials obtained from the same source. Having secured these, we purchased a few frozen salmon, some seals' blubber and venison, but could not prevail upon them to part with more than one of their fine dogs.

One of their sledges was made of two stout pieces of wood, which might have been a boat's keel.

All the old people recollected the visit of the 'Victory.' An old man told me his name was "Ooblooria:" I recollected that Sir James Ross had employed a man of that name as a guide, and reminded him of it; he was, in fact, the same individual, and he inquired after Sir James by his Esquimaux name of "Agglugga."

I inquired after the man who was furnished with a wooden leg by the carpenter of the 'Victory:' no direct answer was given, but his daughter was pointed out to me. Petersen explained to me that they do not like alluding in any way to the dead, and that, as my question was not answered, it was certain the man was no longer amongst the living.

None of these people had seen the whites: one man said he had seen their bones upon the island where they died, but some were buried. Petersen also understood him to say that the boat was crushed by the ice. Almost all of them had part of the plunder; they say they will be here when we return, and will trade more with us; also that we shall find natives upon Montreal Island at the time of our arriving there.

Next morning, 4th March, several natives came to us again. I bought a spear 6½ feet long from a man who told Petersen distinctly that a ship having three masts had been crushed by the ice out in the sea to the west of King William's Island, but that all the people landed safely; he was not one of those who were eye-witnesses of it; the ship sunk, so nothing was obtained by the natives from her; all that they have got, he said, came from the island in the river. The spear staff appears to have been part of the gunwale of a light boat. One old man, "Oo-na-lee," made a rough sketch of the coast-line with his spear upon the snow, and said it was eight journeys to where the ship sank, pointing in the direction of Cape Felix. I can make nothing out of his rude chart.

The information we obtained bears out the principal statements of Dr. Rae, and also accounts for the disappearance of one of the ships; but it gives no clue to the whereabouts of the other, nor the direction whence the ships came. One thing is tolerably certain—the crews did not at any time land upon the Boothian shore.

These Esquimaux were all well clothed in reindeer dresses, and looked clean; they appeared to have abundance of provisions, but scarcely a scrap of wood was seen amongst them which had not come from the lost expedition. Their sledges, with the exception of the one already spoken of, were wretched little affairs, consisting of two frozen rolls of sealskins coated with ice, and attached to each other by bones, which served as the crossbars. The men were stout, hearty fellows, and the women arrant thieves, but all were goodhumoured and friendly. The women were decidedly plain; in fact, this term would have been flattering to most of them; yet there was a degree of vivacity and gentleness in the manners of some that soon reconciled us to these Arctic specimens of the fair sex. They had fine eyes and teeth, as well as very small hands, and the young girls had a fresh rosy hue not often seen in combination with olive complexions.

Esquimaux mothers carry their infants on their backs within their large fur dresses, and where the babes can only be got at by pulling them out over the shoulder. Whilst intent upon my bargaining for silver spoons and forks belonging to Franklin's expedition, at the rate of a few needles or a knife for each relic, one pertinacious old dame, after having obtained all she was likely to get from me for herself, pulled out her infant by the arm, and quietly held the poor little creature (for it was perfectly naked) before me in the breeze, the temperature at the time being 60° below freezing point! Petersen informed me that she was begging for a needle for her child. I need not say I gave it one as expeditiously as possible; yet sufficient time elapsed before the infant was again put out of sight to alarm me considerably for its safety in such a temperature. The natives, however, seemed to think nothing of what looked to me like cruel exposure of a naked baby.

We now returned to the ship with all the speed we could command; but stormy weather occasioned two days' delay, so that we did not arrive on board until the 14th March. Though considerably reduced in flesh, I and my companions were in excellent health, and blessed with insatiable appetites. On washing our faces, which had become perfectly black from the soot of our blubber lamp, sundry scars, relics of frost-bites, appeared; and the tips of our fingers, from constant frost-bites, had become as callous as if seared with hot iron.

In this journey of twenty-five days we travelled 360 geographical miles (420 English), and completed the discovery of the coast-line of continental America, thereby adding about 120 miles to our charts. The mean temperature throughout the journey was 30° below zero of

Fahrenheit, or 62° below the freezing point of water.

On reaching the ship, I at once assembled my small crew, and told them of the information we had obtained, pointing out that there still remained one of the ships unaccounted for, and therefore it was necessary to carry out all our projected lines of search.

During this journey I acquired the Arctic accomplishment of eating frozen blubber, in delicate little slices, and vastly preferred it to frozen pork. At the present moment I do not think I could even taste it, but the same privation and hunger which induced me to eat of such food would doubtless enable me again to partake of it *very kindly*.

I shot a couple of foxes which came playing about the dogs; conscious of their superior speed, they were very impudent, snapping at the dogs' tails, and passing almost under their noses. I shot these foxes, intending to eat them; but the dogs anticipated me with respect to one; the other we feasted off at our mess-table, and thought it by no means bad; it was insipid, but decidedly better to our tastes than preserved meat.

Captain Allen Young and his party had returned on board on the 3rd of March, having placed their depôt upon the shore of Prince of Wales' Land, about 70 miles S.W. of the ship. Young found the ice in Bellot Strait so rough as to be impassable, and was obliged to adopt the lake route. Prince of Wales' Land was found to be composed of limestone; the shore was low, and fringed for a distance of ten miles to seaward with an ancient land-floe. The remaining width of the strait between this land (North Somerset) and Prince of Wales' Land was about 15 miles, and this space was composed of ice formed since September last; this was the water we looked at so anxiously last autumn from Cape Bird and Pemmican Rock. His party lived in their tent, protected from the wind by snow walls, and, like ourselves, escaped with a few trivial frost-bites. So far all was very satisfactory, the general health good, and the eagerness of my crew to commence travelling quite charming.

Young proposed carrying out another depôt to the north-west, in order to explore well up Peel Strait, and would have started on the 17th, but the weather was too severe. The day was spent in a fruitless search for three casks of sugar—a serious and unaccountable deficiency—but, as it was important to replace them with as little delay as possible, Young set off on the 18th, although it blew a N.W. gale at the time, with two men and eighteen dogs, for Fury Beach; failing to find the requisite quantity there, he will go on to Port Leopold.

CHAPTER XIII.

Dr. Walker's sledge journey — Snow-blindness attacks Young's party — Departure of all sledge-parties — Equipment of sledge-parties — Meet the same party of natives — Intelligence of the second ship — My depôt robbed — Part company from Hobson — Matty Island — Deserted snow-huts — Native sledges — Land on King William Land.

DOCTOR WALKER'S zeal for travelling was not to be restrained; I therefore gladly availed myself of his willingness to go with a party to Cape Airey and bring back the depôt of provisions left there in August last. These trips will delay our spring journeys for a few days.

During my absence from the 'Fox' the weather was often stormy, and temperature unusually low; the mean for the month of February was – 36°, showing it to be one of the coldest on record. When possible the men were allowed to go out shooting, and obtained fifty or sixty ptarmigan and a hare; a few foxes were taken in traps, and two reindeer were seen.

Yesterday two bears came near the ship, but were frightened away by the dogs. Hobson shot three ptarmigan. To-day I rambled over the hills, the weather being fine, and saw a hare.

29*th.*—Continued fine weather. A couple more foxes and a lemming in its *brown* coat have been captured, and a hare and four ptarmigan shot. This fine bright weather seems to have awakened the lemmings and ermines; their tracks, which were very rarely seen during winter, are now tolerably numerous; foxes appear in greater numbers, probably following up the ptarmigan from the south. The thermometer ranges between zero and − 20°; it has once been up to + 13°. When exposed to a noonday sun against the ship's side it rises 50° higher. The earth-thermometer—placed 2 feet 2 inches beneath the surface—which gradually fell until the 10th of this month, has now begun to ascend; its minimum was + ½°; much snow also lay over it, 6 feet deep at this season.

On the 25th Dr. Walker and his party returned, not having been able to find the depôt. They found a barrel of flour upon the beach a few miles south of Brentford Bay; it appeared to have lain there for years, just inside a shingle projection, which kept off the ice pressure, so that it had not been forced up high upon the beach; the ice which bore it there—probably from Port Leopold—had disappeared, and the cask was frozen into the shingle. The heading has been brought on board, but the "scribing" upon it is very indistinct, and unintelligible to us. The flour is of the ordinary description used in the navy, and known as "seconds;" most of it was good, and a plain pudding made of it for our mess could not be distinguished from fresh flour. A specimen has been preserved with the view of identifying it with the Fury Beach or Port Leopold stores of flour. With the exception of a solitary bear, the party saw no living creatures. The shore along which they travelled was a very low shingly limestone.

Last evening I was delighted to see Young and his two dog-sledges heave in sight; he brought about 8 cwt. of sugar from Fury Beach, but not without much difficulty, owing to the roughness of the pack in Creswell Bay, and also to the breaking down of one of his sledges; to avoid this pack he found it necessary to travel nearly all round Creswell Bay. Cape Garry he describes as a gradually-curved extent of flat land, and not the decided cape it appears to be upon the chart; two reindeer were seen near it, and during the journey four bears; no other animals were met with. His labours had been very severe; one sledge broke down and all the sugar had to be piled upon the other:

the consequence was that the sledge was so heavily loaded that it would only run freely after the dogs on smooth ice; and directly any hummocks were encountered, the dogs, with their usual instinct, not to drag a sledge unless it does run freely, would lie down, and oblige Captain Young and his two men to unload and carry the packages, over the obstacle, upon their own backs. After this, snow-blindness came on; Young and one of his men became blind as kittens; and the third man had to load, lead, and unload them, when these portages occurred. Young's Esquimaux dog-driver, Samuel, was quite blind when the party reached the ship. Two dogs, not choosing to allow themselves to be caught and put in harness, had been left behind at the last encampment.

There still remains at Fury Beach an immense stack of preserved vegetables and soups; the party supped off them and found them good. Young brought me back two specimen tins of "carrots plain" and "carrots and gravy." All small casks and packages were covered with snow; of the large ones which appeared through it, he saw thirty-four casks of flour, five of split peas, five of tobacco, and four of sugar. Only a very few tons of coals remained. There were two boats, a short four-oared gig and a large cutter; the former required nothing but caulking to make her serviceable, but the latter had a large portion of one bow and side cut out, as if for making, or repairing flat sledges. No record was found.

We have now enough sugar to last us for seven or eight months, but by the survey of provisions which has just been completed, we find a deficiency of many other articles, including three casks of salt beef. Fortunately this is of no consequence, as we have abundance of both salt and preserved meat, but it shows the alarming extent to which a negligent steward may mislead one. This unfortunate man has now got scurvy; want of exercise and fresh air is the apparent cause, combined with irregular living; the spirits have hitherto been in his charge.

The bustle of preparation for the extended searching journeys has been exciting. Hobson's party and my own are now all prepared, and Young having returned, we purpose setting out on the 2nd April—God willing. Young's new sledge will be ready, and he will also start a few days after us. All our winter defences of snow, our porches, our deck-layer, and our external embankment, have been removed. Dr. Walker, of necessity, remains in charge of the ship, with two stewards, a cook, a carpenter, and a stoker. My party, as well as Hobson's, will be provisioned, including the depôts, for an absence of about eighty-four days; but not being able to afford auxiliary or supporting sledge parties, much time will be occupied in transporting our depôts further out, in order that we may start with as much as we can possibly carry, from the Magnetic Pole, besides leaving there a depôt for our return.

The declinometer was taken on board two days ago; hourly observations have been made with it for more than five months: we can no longer spare any one for this interesting duty.

* * * * *

24th June.—One thing is certain, the wild sort of tent-life we lead in Arctic exploration quite unfits one for such tame work as writing up a journal; my present attempt will illustrate the fact,—yet with such ample materials what a deeply interesting volume might be written! Since I last opened this familiar old diary—the repository alike of dry facts and the most trivial notes—winter has passed away, summer is far advanced, and the glorious sun is again returning southward. We too have endeavoured to move on with the times and seasons.

As for myself—I have visited Montreal Island,

completed the exploration and circuit of King William's Island, passing on foot through the only feasible North-West Passage; but all this is as nothing to the interest attached to the *Franklin records* picked up by Hobson, and now safe in my possession! We now know the fate of the 'Erebus' and 'Terror.' The sole object of our voyage has at length been completed, and we anxiously await the time when escape from these bleak regions will become practicable.

* * * * *

The morning of April 2nd was inauspicious, but as the day advanced the weather improved, so that Hobson and I were able to set out upon our journeys; we each had a sledge drawn by four men, besides a dog-sledge, and dog-driver. Mr. Petersen having volunteered his services to drive my dogs,—an offer too valuable to be declined,—managed my dog-sledge throughout. Our five starveling puppies were harnessed, for the first time in their lives, to a small sledge which I drove myself, intending to sell them to the Esquimaux, if I could get them to drag their own supply of provisions so far. The procession looked imposing—it certainly was deeply interesting; there were five sledges, twelve men, and seventeen dogs, the latter of all sizes and shapes. The ship hoisted the Royal Harwich Yacht flag, and our sledges displayed their gay silk banners; mine was a very beautiful one, given me by Lady Franklin; it bears her name in white letters upon a red ground, and is margined with white embroidery; it was worked by the sisters of Captain Collinson.

The equipment of my sledge-party and the weights were as follows: those of Hobson and Young were almost precisely similar.

	lbs. weight.
Two sledges and fittings complete	110
Tent, waterproof blanket, floorcloth, two sleeping-robes, and six blanket sleeping-bags	90
Cooking-utensils, shovel, saw, snow-knife, and sundry small articles	40
Sledge-gun and ammunition	20
Magnetic and astronomical instruments	60
Six knapsacks, containing spare clothing	60
Various tins and bags, in which provision and fuel were stored	50
Articles for barter	40
Provisions	930
Total	1400

The load for each man to drag was fixed at 200 lbs., and for each dog 100 lbs. Our provisions consisted mainly of pemmican, biscuit, and tea, with a small addition of boiled pork, rum, and some tobacco.

The men being untrained to the work, and sledges heavily laden, our march was fatiguing and slow. We encamped that night upon the long lake. On the second day we reached the western sea, and upon the third, aided by our sledge sails, we advanced some miles beyond Arcedeckne Island.

The various depôts carried out with so much difficulty and danger in the autumn, were now gathered up as we advanced, until at length we were so loaded as to be compelled to proceed with one-half at a time, going three times over the same ground. For six days this tedious mode of progression was persevered in, by which time (15th April) we reached the low limestone shore in latitude 71° 7′ N., and which continues thence in almost a straight line southward for 60 or 70 miles. We now commenced laying down provisions for our consumption upon the return journey; and the snow being unusually level, we were able to advance with the whole of our remaining provisions, amounting to nearly sixty days' allowance.

Hitherto the temperature continued low, often nearly 30° below zero, and at times with cutting north winds, bright sun, and intensely strong snow glare. Although we wore coloured spectacles, yet almost all suffered great inconvenience and considerable pain from inflamed eyes. Our faces were blistered, lips and hands cracked,—never were men more disfigured by the combined effects of bright sun and bitterly cold winds; fortunately no serious frost-bites occurred, but frost-bitten faces and fingers were universal.

On 20th April, in latitude 70½° N., we met two families of natives, comprising twelve individuals; their snow huts were upon the ice three-quarters of a mile off shore, and their occupation was seal-hunting. They were the same people with whom I had communicated at Cape Victoria in February.

Old Oo-na-lee laid his hands on Petersen's shoulders to measure their width, and said, "He is fatter now:" true enough, the February temperature and sharp marching had caused us both at that time to shrink considerably.

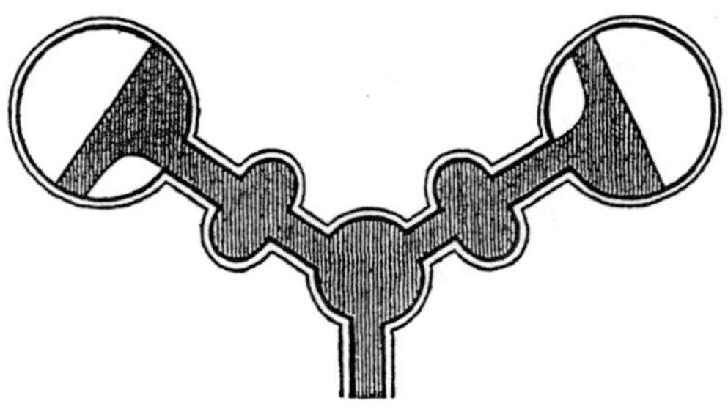

Their snow huts were built in the above form, the common entrance and both passages being just sufficiently high to get in without having to crawl upon our hands and knees. A slab of ice in the roof admitted sufficient light. A snow bank or bench two feet high, and occupying half the area of each hut, was covered with reindeer skins, and formed the family place of repose. An angular snow bench served as the kitchen table, and immediately beside it sat the lady of the establishment attending the stone lamp which stood thereon, and the stone cooking vessel suspended over it. The lamp was a shallow open vessel, the fuel seal oil, and the wick dried moss. Her "tinder box" was a little seal-skin bag of soft dry moss, and with a lump of iron pyrites and a broken file she struck fire upon it. I purchased the file because it was marked with the Government broad arrow.

We saw two large snow shovels made of mahogany board, some long spear handles, a bow of English wood, two preserved meat tins, and a deal case which might have once contained a large telescope or a barometer; it measured 3 feet 1 inch in length by 9 inches wide and 3½ inches deep; there was no lid, but part of the brass hinges remained.

I also purchased a knife which had some indistinct markings on it such as ship's cutlasses or swords usually have; the man told us it had been picked up on the shore near where a ship lay stranded; that it was then about the length of his arm, but his countryman who picked it up broke it into lengths to make knives.

After much anxious inquiry we learned that two ships had been seen by the natives of King William's Island; one of them was seen to sink in deep water, and nothing was obtained from her, a circumstance at which they expressed much regret; but the other was forced on shore by the ice, where they suppose she still remains, but is much broken. From this ship they have obtained most of their wood, &c.; and Oot-loo-lik is the name of the place where she grounded.

Formerly many natives lived there, now very few remain. All the natives have obtained plenty of the wood.

The most of this information was given us by the young man who sold the knife. Old Oo-na-lee, who drew the rough chart for me in March, to show where the ship sank, now answered our questions respecting the one forced on shore; not a syllable about her did he mention on the former occasion, although we asked whether they knew of only one ship? I think he would willingly have kept us in ignorance of a wreck being upon their coasts, and that the young man unwittingly made it known to us.

The latter also told us that the body of a man was found on board the ship; that he must have been a very large man, and had long teeth: this is all he recollected having been told, for he was quite a child at the time.

They both told us it was in the fall of the year—that is, August or September—when the ships were destroyed; that all the white people went away to the "large river," taking a boat or boats with them, and that in the following winter their bones were found there.

These two Esquimaux families had been up as far north as the Tasmania Group * in latitude 71¼° N., and were returning to Nĕitchīllĕe, hunting seals by the way; those we met at Cape Victoria had already gone there. The nearest natives to us at present, they said, were residing at the island of Amitoke, ten days' journey distant from here. Can this Amitoke be Matty Island?

We purchased some seal's blubber and flesh, as well as their two only dogs; but next morning Oo-na-lee repented his bargain, or feigned to do so, but as he came without the knife to exchange

* These islands were so named by me, at the request of Lady Franklin, in grateful acknowledgment of many proofs of affectionate sympathy received from the colony over which her husband presided for several years, and, in particular, of the large contributions raised there in aid of her expeditions of search.

back we retained his dog; he tried to steal a tin vessel off one of the sledges, and perhaps it was for the purpose of regaining our favour that he made known to us, just as we were starting, that his countrymen had followed my homeward track in March, discovering my depôt of blubber, articles for barter, and two revolvers, and carried them all off to Nĕitchīllĕe,—by no means pleasant intelligence; their dogs must have enabled them to find the blubber by scenting it, for it was buried under 4 feet of snow, and strong winds obliterated all traces upon the surface.

I was now glad we had purchased both the dogs of the men, as it would probably prevent their seeking for our depôts to the northward; the knowledge of the insecurity of *all* depôts amongst these people will keep us on our guard for the future. I regretted the loss of the pistols, as it left my party with no other arms than two guns.

Oo-na-lee told us when we first met him that one of his countrymen was very sick; not seeing a sick man in their huts, we forgot all about it until after starting, when Petersen interpreted to me Oo-na-lee's parting information, and told me how he described that the breech of the revolver turned round; it then occurred to me that one of the men might have been wounded, —they had discovered how to cock the locks, and the pistols were loaded and capped.

Oo-na-lee was well acquainted with the coast-line up to Bellot Strait, and had names for the different headlands, although he had never been so far north; he made many inquiries about the position of our ship, her size, and the number of men. Had he been able to travel so far with his wife and several young children, and without sledge or dogs, I think he certainly would have gone up to Port Kennedy; we did not give him any encouragement to do so. His wife was one of the most importunate of the many women we saw at Cape Victoria in March. She was the woman who plucked out an infant by its arm from inside her dress, and exposed it regardless of – 30° and a fresh wind, as I have previously told.

The information respecting *both* the missing ships was most important, and it remained for us to discover, if possible, the stranded ship.

Continuing our journey, we crossed a wide bay upon level ice, and the most perfectly smooth hard snow I ever saw; there must have been much open water here late last autumn. Seven or eight snow huts, recently abandoned, were found near the magnetic pole. During the 25th, 26th, and 27th we were confined to our tents by a very heavy south-east gale, with severe cold. Early on the 28th we reached Cape Victoria; here Hobson and I separated. He marched direct for Cape Felix, King William's Land, whilst I kept a more southerly course. Not daring to leave depôts upon this coast, we carried on our whole supply, intending to deposit a small portion upon the Clarence Islands.

Hobson was unwell when we parted, complaining of stiffness and pain in his legs; neither of us then suspected the cause. I gave him directions to search the west coast of King William's Island for the stranded ship and for records, and to act upon such information as he might obtain in this way, or from the natives; but should that shore prove destitute of traces, to carry out if possible our original plan for the completion of discovery and search upon Victoria Land, comprising the blank space between the extremes visited by Captain Collinson and Mr. Wynniatt.

I soon found that my party had to labour across a rough pack; nor was it until the third day that we completed the traverse of the strait, and encamped near to the entrance of Port Parry, in King William's Island. Although the weather was clear, and that by our reckoning we passed directly over the assigned position of the two southern of the Clarence Islands, yet we saw nothing of them.

A day was devoted to securing a depôt in a huge mass of grounded ice, and in repairing and drying equipment, or, to speak more correctly, in getting rid of the ice which encumbered our sleeping bags and gear: this we effected by beating them well and exposing them to the direct rays of the sun. Magnetic and other observations gave me ample employment, the only *immediate* result of which was my being almost snow-blind for the two following days.

On May 2nd we set off again briskly; our load being diminished to thirty days' provisions, and the sledge sail set, we soon reached the land, and travelled along it for Cape Sabine; it was very thick weather, and we were unable to see any distance in consequence of the mist and snowdrift. The following day was no better, and the shore, which we dared not leave to cross the bays, was extremely low.

We soon discovered that we had strayed inland; but, guided by the wind, continued our course. Upon May 4th we descended into Wellington Strait, and the weather being tolerably clear, crossed over to the south-west extreme of Matty Island, in the hope of meeting with natives, no traces of them having been met with since leaving Cape Victoria. Off this south-west point we found a deserted village of nearly twenty snow huts, besides several others, within a few miles upon either side of it; in all of them I found shavings or chips of different kinds of wood from the lost expedition; they appeared to have been abandoned only within a fortnight or three weeks. Abundance of blubber was gathered up to increase our stock of fuel, and, had we encamped here, the dogs would have feasted sumptuously off the scraps and bones of seals strewed about.

The runners (or sides) of some old sledges left here were very ingeniously formed out of rolls of sealskin, about 3½ feet long, and flattened so as to be 2 or 3 inches wide and 5 inches high; the sealskins appeared to have been well soaked and then rolled up, flattened into the required

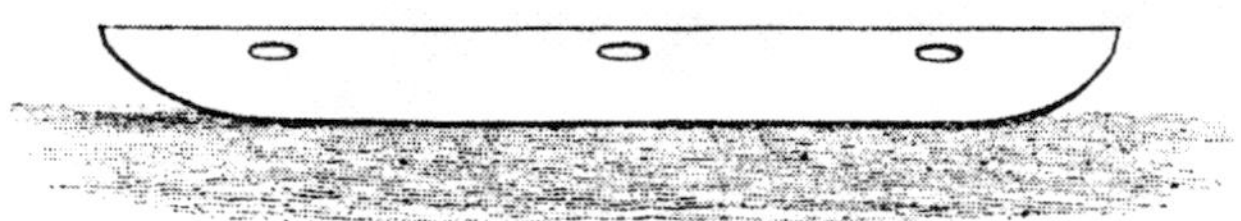

form and allowed to freeze. The underneath part was coated with a mixture of moss and ice laid smoothly on by hand before being allowed to freeze, the moss, I suppose, answering the purpose of hair in mortar, to make the compound adhere more firmly.

From this spot the shore-line of Matty Island turned sharply to the N.N.E.; there were some considerable islands to the east, but thinking the most southerly of this group, named "Owut-tā" by the Esquimaux, the most likely place to find the natives, I pushed on in that direction until we encamped. Thick fog enveloped us for the next two days; we could not find the island, but found a very small islet near it, off which was another snow-village very recently abandoned, the sledge tracks plainly showing that the inhabitants had gone to the E.N.E., which is straight for Nĕitchīllĕe. It was now evident that these places of winter resort were deserted, and that here at least we should not find any natives; I was the more sorry at having missed them, as, from the quantity of wood chips about the huts, they probably had visited the stranded ship alluded to by the last Equimaux we had met, and the route to which lies up an inlet visible from here, and then overland three or four days' journey to the westward, until the opposite coast of King William's Land is reached.

The largest huts measured 12 feet in diameter,

by 6 or 7 feet high; the greater part were constructed in pairs, having a passage 20 or 25 feet long, serving as the common entrance; where the passage divides into two branches, there was a small hut, which served as a sort of antechamber for the reception of such articles as were intended to remain frozen.

CHAPTER XIV.

Meet Esquimaux —News of Franklin's people — Frighten a solitary party — Reach the Great Fish River — On Montreal Island — Total absence of all relics — Examine Ogle Peninsula — Discover a skeleton — Vagueness of Esquimaux information— Cape Herschel — Cairn.

7th May.—To avoid snow-blindness, we commenced night-marching. Crossing over from Matty Island towards the King William Island shore, we continued our march southward until midnight, when we had the good fortune to arrive at an inhabited snow village. We found here ten or twelve huts and thirty or forty natives of King William's Island; I do not think any of them had ever seen white people alive before, but they evidently knew us to be friends. We halted at a little distance, and pitched our tent, the better to secure small articles from being stolen whilst we bartered with them.

I purchased from them six pieces of silver plate, bearing the crests or initials of Franklin, Crozier, Fairholme, and McDonald; they also sold us bows and arrows of English woods, uniform and other buttons, and offered us a heavy sledge made of two short stout pieces of curved wood, which no mere boat could have furnished them with, but this of course we could not take away; the silver spoons and forks were readily sold for four needles each.

They were most obliging and peaceably disposed, but could not resist the temptation to steal, and were importunate to barter everything they possessed; there was not a trace of fear, every countenance was lighted up with joy; even the children were not shy, nor backward either, in crowding about us, and poking in everywhere. One man got hold of our saw, and tried to retain it, holding it behind his back, and presenting his knife in exchange; we might have had some trouble in getting it from him, had not one of my men mistaken his object in presenting the knife towards me, and run out of the tent with a gun in his hand; the saw was instantly returned, and these poor people seemed to think they never could do enough to convince us of their friendliness; they repeatedly tapped me gently on the breast, repeating the words "Kammik toomee" (We are friends).

Having obtained all the relics they possessed, I purchased some seal's flesh, blubber, frozen venison, dried and frozen salmon, and sold some of my puppies. They told us it was five days' journey to the wreck,—one day up the inlet still in sight, and four days overland; this would carry them to the western coast of King William Land; they added that but little now remained of the wreck which was accessible, their countrymen having carried almost everything away. In answer to an inquiry, they said she was without masts; the question gave rise to some laughter amongst them, and they spoke to each other about *fire*, from which Petersen thought they had burnt the masts through close to the deck in order to get them down.

There had been *many books* they said, but all have long ago been destroyed by the weather; the ship was forced on shore in the fall of the year by the ice. She had not been visited during this past winter, and an old woman and a boy were shown to us who were the last to visit the wreck; they said they had been at it during the winter of 1857–8.

Petersen questioned the woman closely, and she seemed anxious to give all the information in her power. She said many of the white men dropped by the way as they went to the Great River; that some were buried and some were not; they did not themselves witness this, but discovered their bodies during the winter following.

We could not arrive at any approximation to the numbers of the white men nor of the years elapsed since they were lost.

This was all the information we could obtain, and it was with great difficulty so much could be gleaned, the dialect being strange to Petersen, and the natives far more inclined to ask questions than to answer them. They assured us we should find natives upon the south shore of King William's Island only three days' journey from here, and also at Montreal Island; moreover they said we might find some at the wreck. For these reasons I did not prolong my stay with them beyond a couple of hours. They seemed to have but little intercourse with other communities, not having heard of our visit to the Boothians two months before: one man even asked Petersen if he had seen his brother, who lived in Boothia, not having heard of him since last summer.

It was quite a relief to get away from these good-humoured, noisy thieves, and rather difficult too, as some of them accompanied us for miles. They had abundance of food, were well clothed, and are a finer race than those who inhabit North Greenland, or Pond's Inlet: the men had their hair cropped short, with the exception of one long, straggling lock hanging down on each side of the face; like the Boothians, the women had lines tattooed upon their cheeks and chins.

We now proceeded round a bay which I named Latrobe in honour of the late Governor of Victoria, and of his brother the head of the Moravian Church in London, both esteemed friends of Franklin.

Finding the "Mathison Island" of Rae to be a flat-topped hill, we crossed over low land to the west of it, and upon the morning of the 10th May reached a single snow hut off Point Booth. I was quite astonished at the number of poles and various articles of wood lying about it, also at the huge pile of walrus' and reindeer's flesh, seal's blubber, and skins of various sorts. We had abundance of leisure to examine these exterior articles before the inmates would venture out; they were evidently much alarmed by our sudden appearance.

A remarkably fine old dog was tied at the entrance—the line being made fast within the long passage—and although he wagged his tail, and received us as old acquaintances, we did not like to attempt an entrance. At length an old man and an old woman appeared; they trembled with fear, and could not, or would not, say anything except "Kammik toomee:" we tried every means of allaying their fears, but their wits seemed paralyzed, and we could get no information. We asked where they got the wood? They purchased it from their countrymen. Did they know the Great River? Yes, but it was a long way off. Were there natives there now? Yes. They even denied all knowledge of white people having died upon their shores. A fine young man came out of the hut, but we could learn nothing of him; they said they had nothing to barter, except what we saw, although we tempted them by displaying our store of knives and needles.

The wind was strong and fair, and the morning intensely cold, and as I could not hope to overcome the fears of these poor people without encamping, and staying perhaps a day with them, I determined to push on, and presented the old lady with a needle as a parting gift.

The principal articles which caught my attention here were eight or ten fir poles, varying in length from 5 to 10 feet, and up to 2½ inches in diameter (these were converted into spear handles and tent poles), a kayak paddle constructed out of the blades of two ash oars, and two large snow shovels 4 feet long, made of thin plank, painted white or pale yellow; these might have been the bottom boards of a boat. There were many smaller articles of wood.

Half a mile further on we found seven or eight deserted snow huts. Bad weather had now fairly set in, accompanied by a most unseasonable degree of cold. On the morning of the 12th May we crossed Point Ogle, and encamped upon the ice in the Great Fish River the same evening; the cold, and the darkness of our more southern latitude, having obliged us to return to day-travelling. All the 13th we were imprisoned in our tent by a most furious gale, nor was it until late on the morning of the 14th that we could proceed; that evening we encamped 2 miles from some small islands which lie off the north end of Montreal Island.

On the morning of the 15th we made only a short march of 6 miles, as one of the men suffered severely from snow-blindness, and I was anxious to recommence night-travelling; encamped in a little bay upon the N.E. side of Montreal Island. The same evening we again set out, although it was blowing very strongly, and "snowing for a wager," as the men expressed it, but it was only necessary for us to keep close along the shore of the island: we discovered, however, a narrow and crooked channel which led us through to the west side of the island, and, one of the men appearing seriously ill, we encamped about midnight.

Whilst encamped this day, explorations were made about the N.E. quarter of the island; islets and rocks were seen to abound in all directions; eventually it proved to be a separate island upon which we had encamped. The only traces or relics of Europeans found were the following articles, discovered by Petersen, beside a native mark (one large stone set upright on the top of another), at the east side of the main—or Montreal—island:—A piece of a preserved meat tin, two pieces of iron hoop, some scraps of copper, and an iron-hoop bolt. These probably are part of the plunder obtained from the boat, and were left here until a more favourable opportunity should offer, or perhaps necessity should compel the depositor to return for them.

All the 16th we were unable to move, not only because Hampton was ill, but the weather was extremely bad, and snow thickly falling with temperature at zero; certainly strange weather for the middle of May! We have not had a single clear day since the 1st of the month.

On the 17th the weather, though dull, was clear, so Mr. Petersen, Thompson, and I set off with the dog-sledge to complete the examination of Montreal Island, leaving the other three men with the tent: we also hoped to find natives, but had not seen any recent traces of them since passing Point Booth. Petersen drove the dog-sledge close along shore round the island to the south, and as far up the east side as to meet our previously explored portion of it, whilst Thompson and I walked along on the land, the one close down to the beach, and the other higher up, examining the more conspicuous parts: in this order we traversed the remaining portion of the island.

Although the snow served to conceal from us any traces which might exist in hollows or sheltered situations, yet it rendered all objects intended to serve as marks proportionably conspicuous; and we may remember that it was in its winter garb that the retreating crews saw Montreal Island, precisely as we ourselves saw it. The island was almost covered with native marks, usually of one stone standing upright upon another, sometimes consisting of three stones, but very rarely of a greater number.

No trace of a cairn could be found.

In examining, with pickaxe and shovel, a collection of stones which appeared to be arranged artificially, we found a quantity of seal's blubber buried beneath; this old Esquimaux câche was near the S.E. point of the island. The interior of the island and the principal islets adjacent were also examined without success, nor was there the slightest evidence of natives having been here during the winter: it is not to be wondered at that we returned in the evening to our tent somewhat dispirited. The total absence of natives was a bitter disappointment; circles of stones, indicating the sites of their tenting places in summer, were common enough.

Montreal Island is of primary rock, chiefly grey gneiss, traversed with whitish vertical bands in a N. and S. direction (by them I often directed my route when crossing the island). It is of considerable elevation, and extremely rugged. The low beaches and grassy hollows were covered with a foot or two of hard snow, whilst all the level, the elevated, or exposed parts were swept perfectly bare; had a cairn, or even a grave, existed (raised as it must be, the earth being frozen hard as rock), we must at once have seen it. If any were constructed they must have been levelled by the natives; every doubtful appearance was examined with the pickaxe.

A remark made by my men struck me as being shrewd; they judged from the washed appearance of the rock upon the east side of

Montreal Island that it must often be exposed to a considerable sea, such as would effectually remove everything not placed far above its reach; when looking over the smooth and frozen expanse one is apt to forget this.

Since our first landing upon King William's Island we have not met with any heavy ice; all along its eastern and southern shore, together with the estuary of this great river, is one vast unbroken sheet formed in the early part of last winter where *no ice previously existed;* this I fancy (from the accounts of Back and Anderson) is unusual, and may have caused the Esquimaux to vary their seal-hunting localities. Mr. Petersen suggested that they might have retired into the various inlets after the seals; and therefore I determined to cross over into Barrow's Inlet as soon as we had examined the Point Ogle Peninsula.

Upon Montreal Island I shot a hare and a brace of willow-grouse. Up to this date we had shot during our journey only one bear and a couple of ptarmigan. The first recent traces of reindeer were met with here.

On the 18th May crossed over to the mainland near Point Duncan, but, Hampton again complaining, I was obliged to encamp. When away from my party, and exploring along the shore towards Elliot Bay, I saw a herd of eight reindeer and succeeded in shooting one of them. In the evening Petersen shot another. Some willow-grouse also were seen. Here we found much more vegetation than upon King William's Island, or any other Arctic land I have yet seen.

On the evening of the 19th we commenced our return journey, but for the three following weeks our route led us over new ground. Hampton being unable to drag, I made over my puppy team to him, and was thus left free to explore and fully examine every doubtful object along our route. I shall not easily forget the trial my patience underwent during the six weeks that I drove that dog-sledge. The leader of my team, named "Omar Pasha," was very willing, but very lame; little "Rose" was coquettish, and fonder of being caressed than whipped, from some cause or other she ceased growing when only a few months old, she was therefore far too small for heavy work; "Darky" and "Missy" were mere pups; and last of all came the two wretched starvelings, reared in the winter, "Foxey" and "Dolly." Each dog had its own harness, formed of strips of canvas, and was attached to the sledge by a single trace 12 feet long. None of them had ever been yoked before, and the amount of cunning and perversity they displayed to avoid both the whip and the work, was quite astonishing. They bit through their traces, and hid away under the sledge, or leaped over one another's backs, so as to get into the middle of the team out of the way of my whip, until the traces became plaited up, and the dogs were almost knotted together; the consequence was I had to halt every few minutes, pull off my mitts, and, at the risk of frozen fingers, disentangle the lines. I persevered, however, and, without breaking any of their bones, succeeded in getting a surprising amount of work out of them. Hobson drove his own dog-sledge likewise, and as long as we were together we helped each other out of difficulties, and they were frequently occurring, for, apart from those I have above mentioned, directly a dog-sledge is stopped by a hummock, or sticks fast in deep snow, the dogs, instead of exerting themselves, lie down, looking perfectly delighted at the circumstance, and the driver has to extricate the sledge with a hearty one, two, three haul! and apply a little gentle persuasion to set his canine team in motion again.

Having searched the east shore of this land for 7 or 8 miles further north, we crossed over into Barrow's Inlet, and spent a day in its examination, but not a trace of natives was met with.

Regaining the shore of Dease and Simpson's Strait, some miles to the west of Point Richardson, we crossed over to King William's Island upon the morning of the 24th, striking in upon it a short distance west of the Peffer River. The south coast was closely examined as we marched along towards Cape Herschel. Upon a conspicuous point, to the westward of Point Gladman, a cairn nearly five feet high was seen, which, although it did not appear to be a recent construction, was taken down, stone by stone, and carefully examined, the ground beneath being broken up with the pickaxe, but nothing was discovered.

The ground about it was much exposed to the winds, and consequently devoid of snow, so that no trace could have escaped us. Simpson does not mention having landed here, or anywhere upon the island except at Cape Herschel, yet it seemed to me strange that natives should construct such a mark here, since a huge boulder, which would equally serve their purpose, stood upon the same elevation, and within a couple of hundred yards. We had previously examined a similar but smaller cairn, a few miles to the eastward.

We were now upon the shore along which the retreating crews must have marched. My sledges of course travelled upon the sea-ice close along the shore; and, although the depth of snow which covered the beach deprived us of almost every hope, yet we kept a very sharp look-out for traces, nor were we unsuccessful. Shortly after midnight of the 25th May, when slowly walking along a gravel ridge near the beach, which the winds kept partially bare of snow, I came upon a human skeleton, partly exposed, with here and there a few fragments of clothing appearing through the snow. The skeleton—now perfectly bleached—was lying upon its face, the limbs and smaller bones either dissevered or gnawed away by small animals.

A most careful examination of the spot was of course made, the snow removed, and every scrap of clothing gathered up. A pocket-book afforded strong grounds for hope that some information might be subsequently obtained respecting the unfortunate owner and the calamitous march of the lost crews, but at the time it was frozen hard. The substance of that which we gleaned upon the spot may thus be summed up :—

This victim was a young man, slightly built, and perhaps above the common height; the dress appeared to be that of a steward or officer's servant, the loose bow-knot in which his neck-handkerchief was tied not being used by seamen or officers. In every particular the dress confirmed our conjectures as to his rank or office in the late expedition,—the blue jacket with slashed sleeves and braided edging, and the pilot-cloth great-coat with plain covered buttons. We found, also, a clothes-brush near, and a horn pocket-comb. This poor man seems to have selected the bare ridge top, as affording the least tiresome walking, and to have fallen upon his face in the position in which we found him.

It was a melancholy truth that the old woman spoke when she said, "they fell down and died as they walked along."

I do not think the Esquimaux had discovered this skeleton, or they would have carried off the brush and comb: superstition prevents them from disturbing their own dead, but would not keep them from appropriating the property of the white man if in any way useful to them. Dr. Rae obtained a piece of flannel, marked

"F. D. V., 1845," from the Esquimaux of Boothia or Repulse Bay: it had doubtless been a part of poor Des Vœux's garments.

At the time of our interview with the natives of King William's Island, Petersen was inclined to think that the retreat of the crews took place in the fall of the year, some of the men in boats, and others walking along the shore; and as only five bodies are said to have been found upon Montreal Island with the boat, this fact favoured his opinion, because so small a number could not have dragged her there over the ice, although they could very easily have taken her there by water. Subsequently this opinion proved erroneous. I mention it because it shows how vague our information was—indeed all Esquimaux accounts are naturally so—and how entirely we were dependent upon our own exertions for bringing to light the mystery of their fate.

The information obtained by Dr. Rae was mainly derived second-hand from the Fish River Esquimaux, and should not be confounded with that received by us from the King William's Island Esquimaux. These people told us they did not find the bodies of the white men (that is, they did not know any had died upon the march) until the following

M'CLINTOCK'S TRAVELLING PARTY DISCOVERING THE REMAINS OF CAIRN AT CAPE HERSCHEL.

Drawn by Captain May.

winter. This is probably true, as it is only in winter and early spring they can travel overland to the west shore, or that they make a practice of wandering along the shore in search of seals and bears.

The remains of those who died in the Fish River may very probably have been discovered in the summer shortly after their decease.

Along the south coast of King William's Land, as upon the mainland, I was sadly disappointed in my expectation of meeting natives. We found only six or eight deserted snow huts, showing that they had recently been here, and consequently there was the less chance of meeting with them on our further progress, as the season had now arrived when they seek the rivers and the favourite haunts and passes of the reindeer in their northern migration.

Hobson was however upon the western coast, and I hoped to find a note left for me at Cape Herschel containing some piece of good news. After minutely examining the intervening coast-line, it was with strong and reasonable hope I ascended the slope which is crowned by Simpson's conspicuous cairn. This summit of Cape Herschel is perhaps 150 feet high, and about a quarter of a mile within the low stony point which projects from it, and on which there was considerable ice pressure and a few hummocks heaped up, the first we had seen for three weeks. Close round this point, or by cutting across it as we did, the retreating parties *must* have passed; and the opportunity afforded by the cairn of depositing in a known position—and that, too, where their own discoveries terminated—some record of their own proceedings, or, it might be, a portion of their scientific journals, would scarcely have been disregarded.

Simpson makes no mention of having left a record in this cairn, nor would Franklin's people have taken any trouble to find it if he had left one; but what now remained of this once "ponderous cairn" was only four feet high; the south side had been pulled down and the central stones removed, as if by persons seeking for something deposited beneath. After removing the snow with which it was filled, and a few loose stones, the men laid bare a large slab of limestone: with difficulty this was removed, then a second, and also a third slab, when they came to the ground. For some time we persevered with a pickaxe in breaking up the frozen earth, but nothing whatever was found, nor any trace of European visitors in its vicinity. There were many old câches and low stone walls, such as natives would use to lurk behind for the purpose of shooting reindeer; and we noticed some recent tracks of those animals which had crossed direct hither from the mainland.

CHAPTER XV.

The cairn found empty—Discover Hobson's letter—Discovery of Crozier's record—The deserted boat—Articles discovered about the boat—The skeletons and relics—The boat belonged to the 'Erebus'—Conjectures.

As the Esquimaux of this land, as well as those of Boothia and Pond's Inlet, have long since given up the practice of building stone dwellings—passing their winters in snow huts and summers in tents—no other traces of them than those described remain; so that when or in what numbers they may have been here one cannot form any opinion, the same câches and hiding-places serving for generations.

I cannot divest myself of the belief that *some record was left here* by the retreating crews, and perhaps some most valuable documents which their slow progress and fast failing strength would have assured them could not be carried much further. If any such were left they have been discovered by the natives, and carried off, or thrown away as worthless.

Doubtless the natives, when they ascertained that famine and fatigue had caused many of the white men "to fall down and die" upon their fearful march, and heard, as they might have done, of its fatal termination upon the mainland, lost no time in following up their traces, examining every spot where they halted, every mark they put up, or stone displaced.

It is easy to tell whether a cairn has been put up or touched within a moderate period of years; if very old, the outer stones have a weathered appearance, lichens will have grown upon the sheltered portions and moss in the crevices; but if recently disturbed, even if a single stone is turned upside down, these appearances are altered. If a cairn has been recently built it will be evident, because the stones picked up from the neighbourhood would be bleached on top by the exposure of centuries, whilst underneath they would be coloured by the soil in which they were imbedded. To the eye of the native hunter these marks of a recent cairn are at once apparent; and unless Simpson's cairn (built in 1839) had been disturbed by Crozier, I do not think the Esquimaux would have been at the trouble of pulling it down to plunder the câche; but, having commenced to do so, would not have left any of it standing, *unless they found what they sought.*

I noticed with great care the appearance of the stones, and came to the conclusion that the cairn itself was of old date, and had been erected many years ago, and that it was reduced to the state in which we found it by people having broken down one side of it, the displaced stones, from being turned over, looking far more fresh than those in that portion of the cairn which had been left standing. It was with a feeling of deep regret and much disappointment that I left this spot without finding some certain record of those martyrs to their country's fame. Perhaps in all the wide world there will be few spots more hallowed in the recollection of English seamen than this cairn on Cape Herschel.

A few miles beyond Cape Herschel the land becomes very low; many islets and shingle-ridges lie far off the coast; and as we advanced we met with hummocks of unusually heavy ice, showing plainly that we were now travelling upon a far more exposed part of the coast-line. We were approaching a spot where a revelation of intense interest was awaiting me.

About 12 miles from Cape Herschel I found

a small cairn built by Hobson's party, and containing a note for me. He had reached this, his extreme point, six days previously, without having seen anything of the wreck, or of natives, but he had found a record—the record so ardently sought for of the Franklin Expedition—at Point Victory, on the N.W. coast of King William's Land.

That record is indeed a sad and touching relic of our lost friends, and, to simplify its contents, I will point out separately the double story it so briefly tells. In the first place, the record paper was one of the printed forms usually supplied to discovery ships for the purpose of being enclosed in bottles and thrown overboard at sea, in order to ascertain the set of the currents, blanks being left for the date and position; any person finding one of these records is requested to forward it to the Secretary of the Admiralty, with a note of time and place; and this request is printed upon it in six different languages. Upon it was written, apparently by Lieutenant Gore, as follows:—

"28 of May, 1847. } H. M. ships 'Erebus' and 'Terror' wintered in the ice in lat. 70° 05′ N., long. 98° 23′ W.

Having wintered in 1846-7 at Beechey Island, in lat. 74° 43′ 28″ N., long. 91° 39′ 15″ W., after having

H. M. S.hips Erebus and Terror
Wintered in the Ice in

28 of May 1847 Lat. 70°5' N Long. 98°23' W

Having wintered in 1846–7 at Beechey Island in Lat 74°43'28" N. Long 91°39'15" W after having ascended Wellington Channel to Lat 77° – and returned by the West side of Cornwallis Island.

Commander.

Sir John Franklin commanding the Expedition.

All well

WHOEVER finds this paper is requested to forward it to the Secretary of the Admiralty, London, *with a note of the time and place at which it was found*: or, if more convenient, to deliver it for that purpose to the British Consul at the nearest Port.

QUINCONQUE trouvera ce papier est prié d'y marquer le tems et lieu ou il l'aura trouvé, et de le faire parvenir au plutot au Secretaire de l'Amirauté Britannique à Londres.

CUALQUIERA que hallare este Papel, se le suplica de enviarlo al Secretario del Almirantazgo, en Londrés, con una nota del tiempo y del lugar en donde se halló.

EEN ieder die dit Papier mogt vinden, wordt hiermede verzogt, om het zelve, ten spoedigste, te willen zenden aan den Heer Minister van de Marine der Nederlanden in 's Gravenhage, of wel aan den Secretaris der Britsche Admiraliteit, te London, en daar by te voegen eene Nota, inhoudende de tyd en de plaats alwaar dit Papier is gevonden geworden.

FINDEREN af dette Papiir ombedes, naar Leilighed gives, at sende samme til Admiralitets Secretairen i London, eller nærmeste Embedsmand i Danmark, Norge, eller Sverrig. Tiden og Stædit hvor dette er fundet önskes venskabeligt paategnet.

WER diesen Zettel findet, wird hier-durch ersucht denselben an den Secretair des Admiralitets in London einzusenden, mit gefälliger angabe an welchen ort und zu welcher zeit er gefundet worden ist.

Party consisting of 2 Officers and 6 Men left the Ships on Monday 24th May 1847

G.M. Gore Lieut.

Chas. F. Des Voeux Mate.

[Left margin, rotated:] 25 April 1848 HM Ships Terror and Erebus were deserted on the 22nd April, 5 leagues NNW of this, having been beset since 12th Sept. 1846. The Officers & Crews consisting of 105 souls – under the Command of Captain F.R.M. Crozier landed here – in Lat. 69°37'42" Long 98°41' This paper was found by Lt. Irving under the cairn supposed to have been

[Right margin, rotated:] built by Sir James Ross in 1831 – 4 miles to the Northward – where it had been deposited by the late Commander Gore in June 1847. Sir James Ross' pillar has not however been found and the paper has been transferred to this position which is that in which Sir J. Ross' pillar was erected – Sir John Franklin died on the 11th June 1847 and the total loss

[Top margin, inverted:] by deaths in the Expedition has been to this date 9 officers & 15 men. James Fitzjames Captain HMS Erebus
F.R.M. Crozier Captain & Senior Offr.
and start on tomorrow 26th for Backs Fish River

London: John Murray, Albemarle Street.

ascended Wellington Channel to lat. 77°, and returned by the west side of Cornwallis Island.

"Sir John Franklin commanding the expedition.

"All well.

"Party consisting of 2 officers and 6 men left the ships on Monday 24th May, 1847.

"GM. GORE, Lieut.
"CHAS. F. DES VŒUX, Mate."

There is an error in the above document, namely, that the 'Erebus' and 'Terror' wintered at Beechey Island in 1846–7,—the correct dates should have been 1845–6; a glance at the date at the top and bottom of the record proves this, but in all other respects the tale is told in as few words as possible of their wonderful success up to that date, May, 1847.

We find that, after the last intelligence of Sir John Franklin was received by us (bearing date of July, 1845) from the whalers in Melville Bay, his Expedition passed on to Lancaster Sound, and entered Wellington Channel, of which the southern entrance had been discovered by Sir Edward Parry in 1819. The 'Erebus' and 'Terror' sailed up that strait for one hundred and fifty miles, and reached in the autumn of 1845 the same latitude as was attained eight years subsequently by H.M.S. 'Assistance' and 'Pioneer.' Whether Franklin intended to pursue this northern course, and was only stopped by ice in that latitude of 77° north, or purposely relinquished a route which seemed to lead away from the known seas off the coast of America, must be a matter of opinion; but this the document assures us of, that Sir John Franklin's Expedition, having accomplished this examination, returned southward from latitude 77° north, which is at the head of Wellington Channel, and re-entered Barrow's Strait by a new channel between Bathurst and Cornwallis Islands.

Seldom has such an amount of success been accorded to an Arctic navigator in a single season, and when the 'Erebus' and 'Terror' were secured at Beechey Island for the coming winter of 1845–6, the results of their first year's labour must have been most cheering. These results were the exploration of Wellington and Queen's Channel, and the addition to our charts of the extensive lands on either hand. In 1846 they proceeded to the south-west, and eventually reached within twelve miles of the north extreme of King William's Land, when their progress was arrested by the approaching winter of 1846–7. That winter appears to have passed without any serious loss of life; and when in the spring Lieutenant Gore leaves with a party for some especial purpose, and very probably to connect the unknown coast-line of King William's Land between Point Victory and Cape Herschel, those on board the 'Erebus' and 'Terror' were "all well," and the gallant Franklin still commanded.

But, alas! round the margin of the paper upon which Lieutenant Gore in 1847 wrote those words of hope and promise, another hand had subsequently written the following words :—

"April 25, 1848.—H. M. ships 'Terror' and 'Erebus' were deserted on the 22nd April, 5 leagues N.N.W. of this, having been beset since 12th September, 1846. The officers and crews, consisting of 105 souls, under the command of Captain F. R. M. Crozier, landed here in lat. 69° 37′ 42″ N., long. 98° 41′ W. Sir John Franklin died on the 11th June, 1847; and the total loss by deaths in the expedition has been to this date 9 officers and 15 men.

(Signed)	(Signed)
"F. R. M. CROZIER,	"JAMES FITZJAMES,
"Captain and Senior Officer.	"Captain H. M. S. Erebus.

"and start (on) to-morrow, 26th, for
Back's Fish River."

This marginal information was evidently written by Captain Fitzjames, excepting only the note stating when and where they were going, which was added by Captain Crozier.

There is some additional marginal information relative to the transfer of the document to its present position (viz., the site of Sir James Ross's pillar) from a spot four miles to the northward, near Point Victory, where it had been originally deposited by the *late* Commander Gore. This little word *late* shows us that he too, within the twelvemonth, had passed away.

In the short space of twelve months how mournful had become the history of Franklin's expedition; how changed from the cheerful "All well" of Graham Gore! The spring of 1847 found them within 90 miles of the known sea off the coast of America; and to men who had already in two seasons sailed over 500 miles of previously unexplored waters, how confident must they then have felt that that forthcoming navigable season of 1847 would see their ships pass over so short an intervening space! It was ruled otherwise. Within a month after Lieutenant Gore placed the record on Point Victory, the much-loved leader of the expedition, Sir John Franklin, was dead; and the following spring found Captain Crozier, upon whom the command had devolved, at King William's Land, endeavouring to save his starving men, 105 souls in all, from a terrible death by retreating to the Hudson Bay territories up the Back or Great Fish River.

A sad tale was never told in fewer words. There is something deeply touching in their extreme simplicity, and they show in the strongest manner that both the leaders of this retreating party were actuated by the loftiest sense of duty, and met with calmness and decision the fearful alternative of a last bold struggle for life, rather than perish without effort on board their ships; for we well know that the 'Erebus' and 'Terror' were only provisioned up to July, 1848.

Another discrepancy exists in the second part of the record written by Fitzjames. The original number composing the expedition was 138 souls,* and the record states the total loss by deaths to have been 9 officers and 15 men, consequently that 114 officers and men remained; but it also states that 105 only landed under Captain Crozier's command, so that 9 individuals are unaccounted for.

Lieutenant Hobson's note told me that he found quantities of clothing and articles of all kinds lying about the cairn, as if these men, aware that they were retreating for their lives,

---

* See Conclusion, p. 348.

had there abandoned everything which they considered superfluous.

. . . .

Hobson had experienced extremely bad weather—constant gales and fogs—and thought he might have passed the wreck without seeing her; he hoped to be more successful upon his return journey.

Encouraged by this important news, we exerted our utmost vigilance in order that no trace should escape us.

Our provisions were running very short, therefore the three remaining puppies were of necessity shot, and their sledge used for fuel. We were also enabled to lengthen our journeys, as we had very smooth ice to travel over, the off-lying islets keeping the rough pack from pressing in upon the shore.

Upon the 29th of May we reached the western extreme of King William's Island, in lat. 69° 08′ N., and long. 100° 08′ W. I named it after Captain Crozier of the 'Terror,' the gallant leader of that "Forlorn Hope" of which we now just obtained tidings. The coast we marched along was extremely low — a mere series of ridges of limestone shingle, almost destitute of fossils. The only tracks of animals seen were those of a bear and a few foxes—the only living creatures a few willow grouse. Traces even of the wandering Esquimaux became much less frequent after leaving Cape Herschel. Here were found only a few circles of stones, the sites of tenting-places, but so moss-grown as to be of great age. The prospect to seaward was not less forbidding—a rugged surface of crushed-up pack, including much heavy ice. In these shallow ice-covered seas, seals are but seldom found; and it is highly probable that all animal life in them is as scarce as upon the land.

From Cape Crozier the coast-line was found to turn sharply away to the eastward; and early in the morning of the 30th May we encamped alongside a large boat—another melancholy relic which Hobson had found and examined a few days before, as his note left here informed me; but he had failed to discover record, journal, pocketbook, or memorandum of any description.

A vast quantity of tattered clothing was lying in her, and this we first examined. Not a single article bore the name of its former owner. The boat was cleared out and carefully swept that nothing might escape us. The snow was then removed from about her, but nothing whatever was found.

This boat measured 28 feet long, and 7 feet

3 inches wide; she was built with a view to lightness and light draught of water, and evidently equipped with the utmost care for the ascent of the Great Fish River; she had neither oars nor rudder, paddles supplying their place; and as a large remnant of light canvas, commonly known as No. 8, was found, and also a small block for reeving a sheet through, I suppose she had been provided with a sail. A sloping canvas roof or rain-awning had also formed part of her equipment. She was fitted with a weather-cloth 9 inches high, battened down all round the gunwale, and supported by 24 iron stanchions, so placed as to serve likewise for rowing thowells. There were 50 fathoms of deep-sea sounding-line near her, as well as an ice grapnel. She appeared to have been originally "carvel" built; but for the purpose of reducing weight, very thin fir planks had been substituted for her seven upper strakes, and put on "clincher" fashion.

The weight of the boat alone was about 700 or 800 lbs. only, but she was mounted upon a sledge of unusual weight and strength. It was constructed of two oak planks 23 feet 4 inches in length, 8 inches in width, and with an average thickness of 2½ inches. These planks formed the sides or runners of the sledge; they were connected by five cross-bars of oak, each 4 feet long, and 4 inches by 3½ inches thick, and bolted down to the runners; the underneath parts of the latter were shod with iron. Upon the cross-bars five saddles or supporting chocks for the boat were lashed, and the drag-ropes by which the crew moved this massive sledge, and the weights upon it, consisted of 2¾-inch whale-line.

I have calculated the weight of this sledge to be 650 lbs.; it could not have been less, and may have been considerably more. The total weight of boat and sledge may be taken at 1400 lbs., which amounts to a heavy load for seven strong healthy men.

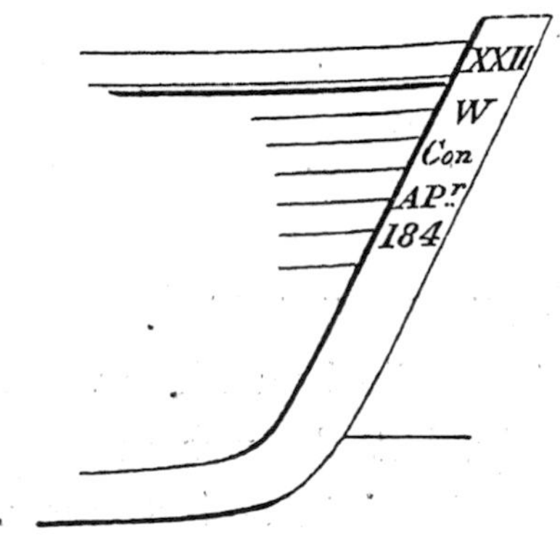

The only markings about the boat were those upon her stem, by which we learned that she was built by contract, was received into Woolwich Dockyard in April, 184 ,* and was numbered 61. There may have been a fourth figure to the right hand, as the stem had been reduced in order to lighten the boat. The ground the sledge rested upon was the usual limestone shingle perfectly flat, and probably overflowed at times every summer, as the stones were embedded in ice.

The boat was partially out of her cradle upon the sledge, and lying in such a position as to lead me to suppose it the effect of a violent north-west gale. She was barely, if at all, above the reach of occasional tides.

One hundred yards from her, upon the land side, lay the stump of a fir-tree 12 feet long, and 16 inches in diameter at 3 feet above the roots. Although the ice had used it roughly during its drift to this shore, and rubbed off every vestige of bark, yet the wood was perfectly sound. It may have been and probably has been lying there for twenty or thirty years, and during such a period would suffer less decay in this region of frost than in one-sixth of the time at home. Within two yards of it I noticed a few scanty tufts of grass.

But all these were after observations; there

* Only the first three figures of the date upon her stem remained, thus—184 .

was that in the boat which transfixed us with awe. It was portions of two human skeletons. One was that of a slight young person; the other of a large, strongly-made, middle-aged man. The former was found in the bow of the boat, but in too much disturbed a state to enable Hobson to judge whether the sufferer had died there; large and powerful animals, probably wolves, had destroyed much of this skeleton, which may have been that of an officer. Near it we found the fragment of a pair of worked slippers, of which I give the pattern, as they may possibly be identified. The lines were white, with a black margin; the spaces white, red, and yellow. They had originally been 11 inches long, lined with calf-skin with the hair left on, and the edges bound with red silk ribbon. Besides these slippers there were a pair of small strong shooting half-boots. The other skeleton was in a somewhat more perfect state,* and was enveloped with clothes and furs; it lay across the boat, under the after-thwart. Close beside it were found five watches; and there were two double-barrelled guns—one barrel in each loaded and cocked—standing muzzle upwards against the

* No part of the skull of either skeleton was found, with the exception only of the lower jaw of each.

boat's side. It may be imagined with what deep interest these sad relics were scrutinised, and how anxiously every fragment of clothing was turned over in search of pockets and pocketbooks, journals, or even names. Five or six small books were found, all of them scriptural or devotional works, except the 'Vicar of Wakefield.' One little book, 'Christian Melodies,' bore an inscription upon the titlepage from the donor to G. G. (Graham Gore?) A small Bible contained numerous marginal notes, and whole passages underlined. Besides these books, the covers of a New Testament and Prayerbook were found.

Amongst an amazing quantity of clothing there were seven or eight pairs of boots of various kinds—cloth winter boots, sea boots, heavy ankle boots, and strong shoes. I noted that there were silk handkerchiefs — black, white, and figured—towels, soap, sponge, toothbrush, and hair-combs; macintosh gun-cover, marked outside with paint A 12, and lined with black cloth. Besides these articles we found twine, nails, saws, files, bristles, wax-ends, sailmakers' palms, powder, bullets, shot, cartridges, wads, leather cartridge-case, knives—clasp and dinner ones — needle and thread cases, slow-match, several bayonet-scabbards cut down into knife-sheaths, two rolls of sheet-lead, and, in short, a quantity of articles of one description and another truly astonishing in variety, and such as, for the most part, modern sledge-travellers in these regions would consider a mere accumulation of dead weight, but slightly useful, and very likely to break down the strength of the sledge-crews.

The only provisions we could find were tea and chocolate; of the former very little remained, but there were nearly 40 pounds of the latter. These articles alone could never support life in such a climate, and we found neither biscuit nor meat of any kind. A portion of tobacco and an empty pemmican-tin, capable of containing 22 pounds weight, were discovered. The tin was marked with an E; it had probably belonged to the 'Erebus.' None of the fuel originally brought from the ships remained in or about the boat, but there was no lack of it, for a drift-tree was lying on the beach close at hand, and had the party been in need of fuel they would have used the paddles and bottom-boards of the boat.

In the after-part of the boat we discovered eleven large spoons, eleven forks, and four tea-spoons, all of silver; of these twenty-six pieces of plate, eight bore Sir John Franklin's crest, the remainder had the crests or initials of nine different officers, with the exception of a single fork which was not marked; of these nine officers, five belonged to the 'Erebus,'—Gore, Le Vesconte, Fairholme, Couch, and Goodsir. Three others belonged to the 'Terror,'—Crozier, (a teaspoon only), Hornby, and Thomas. I do not know to whom the three articles with an owl engraved on them belonged, nor who was the owner of the unmarked fork, but of the owners of those we can identify, the majority belonged to the 'Erebus.' One of the watches bore the crest of Mr. Couch, of the 'Erebus,' and as the pemmican tin also came from that ship, I am inclined to think the boat did also; the authorities at Woolwich could tell (by her number) to which ship she was supplied; and as one of the pocket chronometers found in the boat was marked, "Parkinson and Frodsham 980," and the other, "Arnold 2020," it could also be ascertained to which ship they had been issued.*

Sir John Franklin's plate perhaps was issued to the men for their use, as the only means of saving it; and it seems probable that the officers generally did the same, as not a single iron

* These chronometers, according to the receipts in office, were supplied one to each ship in 1845; but it is impossible to tell to which ship the boat belonged, as the number is imperfect.

spoon, such as sailors always use, has been found. Of the many men, probably twenty or thirty, who were attached to this boat, it seemed most strange that the remains of only two individuals were found, nor were there any graves upon the neighbouring flat land; indeed, bearing in mind the season at which these poor fellows left their ships, it should be remembered that the soil was then frozen hard, and the labour of *cutting* a grave very great indeed.

I was astonished to find that the sledge was directed to the N.E., exactly for the next point of land for which we ourselves were travelling!

The position of this abandoned boat is about 50 miles—as a sledge would travel—from Point Victory, and therefore 65 miles from the position of the ships; also it is 70 miles from the skeleton of the steward, and 150 miles from Montreal Island: it is moreover in the depth of a wide bay, where, by crossing over 10 or 12 miles of very low land, a great saving of distance would be effected, the route by the coast-line being about 40 miles.

A little reflection led me to satisfy my own mind at least, that the boat was returning to the ships: and in no other way can I account for two men having been left in her, than by supposing the party were unable to drag the boat further, and that these two men, not being able to keep pace with their shipmates, were therefore left by them supplied with such provisions as could be spared to last until the return of the others from the ship with a fresh stock.

Whether it was the intention of the retroceding party to await the result of another season in the ships, or to follow the track of the main body to the Great Fish River, is now a matter of conjecture. It seems highly probable that they had purposed revisiting the boat, not only on account of the two men left in charge of it, but also to obtain the chocolate, the five watches, and many other articles which would otherwise scarcely have been left in her.

The same reasons which may be assigned for the return of this detachment from the main body, will also serve to account for their not having come back to their boat. In both instances they appear to have greatly overrated their strength, and the distance they could travel in a given time.

Taking this view of the case, we can understand why their provisions would not last them for anything like the distance they required to travel; and why they would be obliged to send back to the ships for more, first taking from the detached party all provisions they could possibly spare. Whether all or any of the remainder of this detached party ever reached their ships is uncertain; all we know is, that they did not re-visit the boat, and which accounts for the absence of more skeletons in its neighbourhood; and the Esquimaux report that there was no one alive in the ship when she drifted on shore, and that but one human body was found by them on board of her.

After leaving the boat we followed an irregular coast-line to the N. and N.W., up to a very prominent cape, which is probably the extreme of land seen from Point Victory by Sir James Ross, and named by him Point Franklin, which name, as a cape, it still retains.

I need hardly say that throughout the whole of my journey along the shores of King William's Land I caused a most vigilant look-out to be kept to seaward for any appearance of the stranded ship spoken of by the natives; our search was however fruitless in that respect.

CHAPTER XVI.

Errors in Franklin's records—Relics found at the cairn—Reflections on the retreat — Returning homeward — Geological remarks — Difficulties of summer sledging — Arrive on board the 'Fox'— Navigable N.W. passage — Death from scurvy — Anxiety for Captain Young — Young returns safely.

ON the morning of 2nd June we reached Point Victory. Here Hobson's note left for me in the cairn informed me that he had not found the slightest trace either of a wreck anywhere upon the coast, or of natives to the north of Cape Crozier.

Although somewhat short of provisions, I determined to remain a day here in order to examine an opening at the bottom of Back Bay, called so after Sir George Back, by his friend Sir James Ross, and which had not been explored. This proved to be an inlet nearly 13 miles deep, with an average width of 1½ or 2 miles; I drove round it upon the dog sledge, but found no trace of human beings; it was filled with heavy old ice, and was therefore unfavourable for the resort of seals, and consequently of natives also.

The direction of the inlet is to the E.S.E.; we found the land on either side rose as we advanced up it, and attained a considerable elevation, except immediately across its head, where alone it was very low; I have conferred upon it the name of Collinson, after one who will ever be distinguished in connexion with the Franklin search, and who kindly relieved Lady Franklin of much trouble by taking upon himself the financial business of this expedition.

An extensive bay, westward of Cape Herschel, I have named after Captain Washington, the hydrographer, a stedfast supporter of this final search.

All the intermediate coast-line along which the retreating crews performed their fearful march is sacred to their names alone.

Hobson's note informed me of his having found a second record, deposited also by Lieut. Gore in May, 1847, upon the south side of Back Bay, but it afforded no additional information.

It is strange that both these papers state the ships to have wintered in 1846-7 at Beechey Island! So obvious a mistake would hardly have been made had any importance been attached to these documents. They were soldered up in thin tin cylinders, having been filled up on board prior to the departure of the travellers; consequently the day upon which they were *deposited* was not filled in; but already the papers were much damaged by rust,—a very few more years would have rendered them wholly illegible. When the record left at Point Victory was opened to add thereto the supplemental information which gives it its chief value, Captain Fitzjames, as may be concluded by the colour of the ink, filled in the date — 28th — in May, when the record was originally deposited. The cylinder containing this record had not been soldered up again; I suppose they had not the means of doing so; it was found on the ground amongst a few loose stones which had evidently fallen along with it from the top of the cairn. Hobson removed every stone of this cairn down to the ground and rebuilt it.

Brief as these records are, we must needs be contented with them; they are perfect models of official brevity. No log-book could be more provokingly laconic. Yet, that *any record at all* should be deposited after the abandonment of the ships, does not seem to have been intended; and we should feel the more thankful to Captains Crozier and Fitzjames, to whom we are indebted for the invaluable supplement; and our gratitude ought to be all the greater when we remember that the ink had to be thawed, and that writing in a tent during an April day in the Arctic regions is by no means an easy task.

Besides placing a copy of the record taken away by Hobson from the cairn, we both put records of our own in it; and I also buried one under a large stone ten feet true north from it, stating the explorations and discoveries we had made.

A great quantity and variety of things lay strewed about the cairn, such as even in their three days' march from the ships the retreating crews found it impossible to carry further. Amongst these were four heavy sets of boat's cooking stoves, pickaxes, shovels, iron hoops, old canvas, a large single block, about four feet of a copper lightning conductor, long pieces of hollow brass curtain rods, a small case of selected medicines containing about twenty-four phials, the contents in a wonderful state of preservation; a dip circle by Robinson, with two needles, bar magnets, and light horizontal needle all complete, the whole weighing only nine pounds; and even a small sextant engraved with the name of "Frederic Hornby" lying beside the cairn without its case. The coloured eye-shades of the sextant had been taken out, otherwise it was perfect; the moveable screws and such parts as come in contact with the observer's hand were neatly covered with thin leather to prevent frost-bite in severe weather.

The clothing left by the retreating crews of the 'Erebus' and 'Terror' formed a huge heap four feet high; every article was searched, but the pockets were empty, and not one of all these articles was marked,—indeed sailors' warm clothing seldom is. Two canteens, the property of marines, were found, one marked "88 C°. Wm. Hedges," and the other "89 C°. Wm. Hether." A small pannikin made out of a two-pound preserved-meat tin had scratched on it "W. Mark."

When continuing my homeward march, and, as nearly as I could judge, 2½ or 2¾ miles to the north of Point Victory, I saw a few stones placed in line, as if across the head of a tenting place to afford some shelter; here it was I think that Lieutenant Gore deposited the record in May, 1847, which was found in 1848 by Lieutenant Irving, and finally deposited at Point Victory. Some scraps of tin vessels were lying about, but whether they had been left by Sir James Ross's party in May, 1830, or by the Franklin Expedition in 1847 or 1848, is uncertain.*

* It is a remarkable circumstance that when, in 1830, Sir James Ross discovered Point Victory, he named two points of land, then

Here ended my own search for traces of the lost ones. Hobson found two other cairns, and many relics, between this position and Cape Felix. From each place where any trace was discovered the most interesting of the relics were taken away, so that the collection we have made is very considerable.

Of these northern cairns I will write a description when I have received Hobson's account of his journey; but here it is as well to state his opinion, as well as my own, that no part of the coast between Cape Felix and Cape Crozier has been visited by Esquimaux since the fatal march of the lost crews in April, 1848; none of the cairns or numerous articles strewed about—which would be invaluable to the natives—or even the driftwood we noticed, had been touched by them. From this very significant fact it seems quite certain that they had not been discovered by the Esquimaux, whose knowledge of the "white men falling down and dying as they walked along" must be limited to the shore-line southward and eastward of Cape Crozier, and where, of course, no traces were permitted to remain for us to find. It is not probable that

in sight, Cape Franklin and Cape Jane Franklin respectively. Eighteen years afterwards Franklin's ships perished within sight of those headlands.

such fearful mortality would have overtaken them so early in their march as within 80 miles by sledge-route from the abandoned ships —such being their distance from Cape Crozier; nor is it probable that we could have passed the wreck had she existed there, as there are no off-lying islands to prevent a ship drifting in upon the beach; whilst to the southward they are very numerous; so much so that a drifting ship could hardly run the gauntlet between them so as to reach the shore.

The coast from Point Victory northward is considerably higher than that upon which we have been so many days; the sea also is not so shallow, and the ice comes close in; to seaward all was heavy close pack, consisting of all descriptions of ice, but for the most part old and heavy.

From Walls' Bay I crossed overland to the eastern shore, and reached my depôt near the entrance of Port Parry on the 5th June, after an absence of thirty-four days. Hence I purposed travelling alongshore to Cape Sabine, in order to avoid the rough ice which we encountered when crossing direct from Cape Victoria in April, and also hoping to obtain a few more observations for the magnetic inclination.

The weather became foggy as we approached

Prince George's Bay, therefore we were obliged to go well into it before attempting to cross. We gained the land—upon the opposite side, as I supposed—and which would lead us direct to Cape Sabine; but when the weather cleared up we saw a long low island to seaward of us, which puzzled me much. Eventually I found we had discovered a strait leading from Prince George's Bay into Wellington Strait, about 8 miles south of Cape Sabine.

This discovery cost us a day's delay, and was therefore unwelcome, as we were then in daily expectation and dread of the thaw, which renders all travelling so very difficult; and we were still 230 long miles from our ship. In this strait we found a deserted snow village of seventeen huts; one of them was unusually large, its internal diameter being 14 feet. The men soon scraped together enough blubber to supply us with fuel for our homeward march. Strewed about on the ice or in every snow hut were shavings and chips of fresh wood; in one of them I found a child's toy—a miniature sledge—made of wood. No traces of natives were found upon either shore at this place, nor had I met with any since leaving the western coast of the island to the southward of Cape Crozier.

Having passed through nearly to the eastern

ISOLATED ICEBERG

Drawn by P. Skelton, from a Sketch by Captain Allen Young

end of the strait, we cut off some distance by crossing overland, so as to reach the sea-coast 3 or 4 miles southward of Cape Sabine. A few willow grouse, two foxes, and a young reindeer were seen. There was some vegetation upon the land, and animals appeared to resort to this locality in tolerable abundance; the contrast between it and the low, barren shores we had so recently come from was striking indeed!

Nothing can exceed the gloom and desolation of the western coast of King William's Island; Hobson and myself had some considerable experience of it; his sojourn there exceeded a month; its climate seems different from that of the eastern coast; it is more exposed to north-west winds, and the air was almost constantly loaded with chilling fogs. Everywhere upon the shores of the island I noticed boulders of dark gneiss; upon the west coast they were generally small, and of a dark gray colour. About the north part of the island Hobson found a good deal of sandstone, the probable result of ice-drift from Melville Island or Banks Land.

This land gives one the idea of its having risen within a recent geological period from the sea—not suddenly, but at regular intervals; the numerous terraces or beach-marks form long horizontal lines, rising very gradually, and in due proportion as their distance increases from the sea; near the shore they are, of course, most distinct. Upon the west coast some fossils were picked up, chiefly impressions of shells.

King William's Island is for the most part extremely barren, and its surface dotted over with innumerable ponds and lakes. It is not by any means the "land abounding with reindeer and musk oxen" which we expected to find: the natives told us there were none of the latter and very few of the former upon it.

On the 8th June the first ducks and brent geese were seen flying northward. Passing over the extreme point of Cape Victoria, Boothia Land, near which we saw the deserted snow huts of our March acquaintances, and shortly afterwards crossing the mouth of the deep bay to the north of it, in which, sheltered by the island, a ship would find security from ice pressure, and very tolerable winter quarters, we again reached the straight low limestone coast of Boothia Felix.

I was unable to make any delay at the Magnetic Pole, nor could I find a trace of Ross's cairn;* but at each of our encampments along

* This cairn, as well as the one built on Point Victory in 1830, was removed by the natives; fortunately they had not visited Point Victory whilst the Franklin cairn and record remained there, otherwise neither cairn nor record would have remained for us to discover.

the coast the magnetic inclination was carefully observed. Throughout my whole journey I availed myself of every opportunity of obtaining these most interesting observations, often remaining up, after we had encamped for rest, six or seven hours in order to do so; but the instruments supplied for this purpose were not well adapted, and occasioned me a vast deal of labour and loss of time, so as to diminish to almost one-third the results I should otherwise have obtained. Much snow has disappeared off the land; and the ridges or ancient beaches, being the parts most free from snow, showed out strongly in long, dark, horizontal lines, rising above each other until lost to view in the interior. Here and there a few fossil shells and corals were picked up, and four or five willow grouse shot.

13*th June.*—We passed from limestone to granite in lat. 71° 10′ N. Here the land attains to considerable elevation. In the hollows of the dark granite rocks we found abundance of water, and also in a few places upon the sea-ice; it was quite evident that in another day or two the snow would altogether yield to the warmth of summer; birds were now frequently seen.

We discovered a narrow channel to the eastward of the one between the Tasmania Group, through which we had passed with so much difficulty in April; our new channel was covered with smooth ice, and was also much shorter.

At one of our depôts lately visited, a note left by Hobson informed me of his being six days in advance of me, and also of his own serious illness; for many days past he had been unable to walk, and was consequently conveyed upon the sledge; his men were hastening home with all their strength and speed, in order to get him under the Doctor's care. We also were doing our best to push on, lest the bursting out of melting snow from the various ravines should render the ice impassable.

On the 15th the snow upon the ice everywhere yielded to the effects of increased temperature; I was, indeed, most thankful at its having remained firm so long. To make any progress at all after this date was of course a very great labour, requiring the utmost efforts of both the men and the dogs; nor was the freezing mixture through which we trudged by any means agreeable: we were often more than knee-deep in it.

We succeeded in reaching False Strait on the morning of the 18th June, and pitched our tent just as heavy rain began to descend; it lasted throughout the greater part of the day. After travelling a few miles upon the Long Lake, further progress was found to be quite impossible, and we were obliged to haul our sledges up off the flooded ice, and commence a march of 16 or 17 miles overland for the ship. The poor dogs were so tired and sore-footed, that we could not induce them to follow us; they remained about the sledges. After a very fatiguing scramble across the hills and through the snow valleys we were refreshed with a sight of our poor dear lonely little 'Fox,' and arrived on board in time for a late breakfast on 19th June.

With respect to a *navigable* North-West Passage, and to the probability of our having been able last season to make any considerable advance to the southward, had the barrier of ice across the western outlet of Bellot Strait permitted us to reach the open water beyond, I think, judging from what I have since seen of the ice in the Franklin Strait, that the chances were greatly in favour of our reaching Cape Herschel, on the S. side of King William's Land, by passing (as I intended to do) *eastward* of that island.

From Bellot Strait to Cape Victoria we found a mixture of old and new ice, showing the exact proportion of pack and of clear water at the setting in of winter. Once to the southward of the Tasmania Group, I think our chief difficulty would have been overcome; and south of Cape Victoria I doubt whether any further obstruction would have been experienced, as but little, if any, ice remained. The natives told us the ice went away, and left a clear sea every year. As our discoveries show the Victoria Strait to be but little more than 20 miles wide, the ice pressed southward through so narrow a space could hardly have prevented our crossing to Victoria Land, and Cambridge Bay, the wintering place reached by Collinson, from the *west*.

No one who sees that portion of Victoria Strait which lies between King William's Island and Victoria Land, as we saw it, could doubt of there being but one way of getting a ship through it, that way being the *extremely* hazardous one of drifting through in the pack.

The wide channel between Prince of Wales' Land and Victoria Land admits a vast and continuous stream of very heavy ocean-formed ice from the N.W., which presses upon the western face of King William's Island, and chokes up Victoria Strait in the manner I have just described. I do not think the North-West Passage could ever be sailed through by passing westward—that is, to windward—of King William's Island.

If the season was so favourable for navigation as to open the northern part of this western sea* (as, for instance, in 1846, when Sir J. Franklin sailed down it), I think but comparatively little difficulty would be experienced in the more southern portion of it until Victoria Strait was reached. Had Sir John Franklin known that a channel existed eastward of King William's *Land* (so named by Sir John Ross), I do not think he would have risked the besetment of his ships in such very heavy ice to the westward of it; but had he attempted the northwest passage by the *eastern* route, he would probably have carried his ships safely through to Behring's Straits. But Franklin was furnished with charts which indicated no passage to the eastward of King William's Land, and made that land (since discovered by Rae to be an island) a peninsula attached to the continent of North America; and he consequently had but one course open to him, and that the one he adopted.

My own preference for the route by the east

* This channel is now named after the illustrious navigator Admiral Sir John Franklin.

side of the island is founded upon the observations and experience of Rae and Collinson in 1851-2-4. I am of opinion that the barrier of ice off Bellot Strait, some 3 or 4 miles wide, was the only obstacle to our carrying the 'Fox,' according to my original intention, southward to the Great Fish River, passing *east* of King William's Island, and from thence to a wintering position on Victoria Land. Perhaps some future voyager, profiting by the experience so fearfully and fatally acquired by the Franklin expedition, and the observations of Rae, Collinson, and myself, may succeed in carrying his ship through from sea to sea; at least he will be enabled to direct all his efforts in the true and only direction. In the mean time to Franklin must be assigned the earliest discovery of the North-West Passage, though not the actual accomplishment of it, in his ships.*

Saturday, 2nd July.—Upon my arrival on board on the morning of the 19th June, my first inquiries were about Hobson; I found him in a

---

* This will be understood when it is recollected that W. of Simpson Straits or Victoria Land a navigable passage to Behring's Straits is known to exist along the coast of North America. Franklin himself, with his companion Richardson, surveyed by far the greater portion of that distance. Franklin's and Parry's discoveries overlap each other in longitude, and for the last thirty years or more the discovery of the North-West Passage has been reduced to the discovery of a link uniting the two.

worse state than I expected. He reached the ship on the 14th, unable to walk, or even stand without assistance; but already he was beginning to amend, and was in excellent spirits. Christian had shot several ducks, which, with preserved potato, milk, strong ale, and lemon-juice, completed a very respectable dietary for a scurvy-stricken patient. All the rest were tolerably well; slight traces only of scurvy in two or three of the men. The ship was as clean and trim as I could expect, and all had well and cheerfully performed their duties during my absence; hardly any game had been shot, except one bear.

The Doctor now acquainted me with the death of Thomas Blackwell, ship's steward, which occurred only five days previously, and was occasioned by scurvy. This man had scurvy when I left the ship in April, and no means were left untried by the Doctor to promote his recovery and rally his desponding energies; but his mind, unsustained by hope, lost all energy, and at last he had to be forcibly taken upon deck for fresh air. For months past the ship's spirits had been of necessity removed from under his control.

When too late his shipmates made it known that he had a dislike to preserved meats, and had lived the whole winter upon salt pork! He also disliked preserved potato, and would not eat it unless watched, nor would he put on clean clothes, which others in charity prepared for him. Yet his death was somewhat unexpected; he went on deck as usual to walk in the middle of the day, and, when found there, was quite dead. His remains were buried beside those of our late shipmate Mr. Brand.

The news of our success to the southward in tracing the footsteps of the lost expedition greatly revived the spirits of my small crew; we wished only for the safe and speedy return of Young and his party.

Captain Young commenced his spring explorations on the 7th April, with a sledge party of four men, and a second sledge drawn by six dogs under the management of our Greenlander, Samuel; finding in his progress that a channel existed between Prince of Wales' Land and Victoria Land whereby his discovery and search would be lengthened, he sent back one sledge, the tent, and four men to the ship, in order to economise provisions, and for forty days journeyed with one man (George Hobday) and the dogs, encamping in such snow lodges as they were able to build.

This great exposure and fatigue, together with extremely bad weather, and a most difficult coast-line to trace, greatly injured his health; he was compelled to return to the ship on 7th June for medical aid, but purposing at all hazards to renew his explorations almost immediately. Dr. Walker met this determination by a strong protest in writing against his leaving the ship again, his health being quite unequal to it; but after three days Young felt himself somewhat better, and, with a zeal which knew no bounds, set off to complete his branch of the search, taking with him both his sledge parties.

From the Doctor's account I felt most anxious for his return, lest his health, or that of his companions, should receive permanent injury; in fact this was now my only cause of anxiety The season was rather forward here, and advancing with unusual rapidity, rain and wind dissolving the snow and ice; there was much water in Bellot Strait, extending from Half-way Island eastward to the table land, and thence in a narrow lane to Long Island. After a day or two I could perceive a vast improvement in Hobson, and my own four men, with the exception of Hampton who required rest, were in sound health; so also was my companion Petersen. On 24th June Christian shot two small reindeer, which gave us 170 lbs. of meat; a few days before that he shot a seal, which afforded two sumptuous meals for all on board.

The time having elapsed during which Young expected to remain absent, and the difficulties of the transit from the western sea having become greatly increased, I set off early on the 25th June with my four men, intending to visit Pemmican Rock; but failing to come across him there, I resolved to carry on provisions as far as Four River Point, in the hope of meeting with him, and of facilitating his return. To our surprise the water had all drained off the frozen surface of the Long Lake, and it therefore afforded excellent travelling. We found the poor dogs lying quietly beside our sledges; they had attacked the pemmican, and devoured a small quantity which was not secured in tin, also some blubber, some leather straps, and a gull that I had shot for a specimen; but they had not apparently relished the biscuit. Poor dogs! they have a hard life of it in these regions. Even Petersen, who is generally kind and humane, seems to fancy they must have little or no feeling: one of his theories is, that you may knock an Esquimaux dog about the head with any article, however heavy, with perfect impunity to the brutes. One of us upbraided him the other day because he broke his whip-handle over the head of a dog. "*That was nothing at all*," he assured us: some friend of his in Greenland found he could beat his dogs over the head with a heavy hammer,—it stunned them certainly,—but by laying them with their mouths open to the wind, they soon revived, got up, and ran about "*all right*."

We lost no time in giving them a good feed, the first for seven days, yet they did not seem unusually hungry, and soon coiled themselves up to sleep again. Whilst the men and dogs were employed next day in conveying a sledge to the east end of the lake, I walked to Cape Bird to look out for the absent party, but they had not yet returned to Pemmican Rock.

When vainly endeavouring, with felonious intentions, to climb up a steep cliff to the breeding-places of some silvery gulls, I saw and shot a brent goose, seated upon an accessible ledge, and made a prize of four eggs; it seems strange that this bird should have selected so unusual a breeding-place. Many seals were basking on the ice, and the watercourse by which our sledges ascended a week before to the Long Lake was now a strong and rapid stream. A few reindeer were seen.

On the 27th I sent three of the men back to the ship, and with Thompson and the dogs went on to Pemmican Rock, where, to our great joy, we happily met Young and his party, who had but just returned there, after a long and successful journey, the particulars of which I will give hereafter.

Young was greatly reduced in flesh and strength, so much weakened indeed that for the last few days he had travelled on the dog sledge; Harvey—also far from well—could just manage to keep pace with the sledge; his malady was scurvy. Their journies had been very depressing; most dismal weather, low dreary limestone shores devoid of game, and no traces of the lost expedition. The news of our success in the southern journies greatly cheered them. On the following day we were all once more on board, and indulging in such rapid consumption of eatables as only those can do who have been much reduced by long-continued fatigue and exposure to cold. Venison, ducks, beer, and lemon-juice, daily; preserved apples and cranberries three times a-week; and pickled whaleskin—a famous antiscorbutic—*ad libitum* for all who liked it. The weather, which for the last week had been wet, windy, and miserable, now set in fair. The carpenter's hammer, and the men's voices at their work, were new and animating sounds.

CHAPTER XVII.

Signs of release—Dearth of animal life—Owl is good beef—Beat out of winter quarters—Our game-list—Reach Fury Beach—Escape from Regent's Inlet—In Baffin's Bay—Captain Allen Young's journey—Disco; sad disappointment—Part from our Esquimaux friends—Adieu to Greenland—Arrive home.

TO-DAY (*2nd July*) I took a long and delightful walk, but shot only two ducks; Petersen went in another direction, and got nothing; Christian, after toiling all day in his kayak, returned with only two divers and a duck. Lately he has obtained for us several king and long-tailed ducks (no eider ducks have been seen), two red-throated divers, and two brent geese, and caught an ermine in its summer coat. Yesterday one of the men brought on board a trout weighing 2 lbs.; he saw a glaucous gull and a fox disputing for it; the former seems to have killed and brought it to land.

The water now washes the south side of the Fox Islands, and extends to the south point of Long Island. The month of June has been somewhat warmer than usual, its mean temperature being $+35\frac{1}{2}°$.

9th.—The ship has been thoroughly cleaned and restowed, remaining provisions examined, tanks filled with fresh water, 12 tons of stone ballast taken in, and everything brought on board that was landed last autumn. Hobson is the only one upon the sick list; but he is able to walk about and does duty. Very few birds, and only one small seal, have been obtained during the week; an occasional great northern diver is seen, and a rare land bird has been shot. We cannot discover the nests of either ducks or geese, and the breeding cliffs of the gulls being inaccessible, we have not got any eggs. I am a close prisoner at the corner of my table, poring over my observation and angle book, and have at length laid down upon paper the west coast of King William's Land to my satisfaction. Tidal observations are commenced; and the aneroid and mercurial barometers are again being compared in order to verify the former.

16*th. Saturday night.*—We are now almost ready for sea. There is a much larger space of water in Bellot Strait, reaching within 300 or 400 yards of us. Long cracks or lanes of water have been seen in Prince Regent's Inlet. The decay of the ice continues, though not with equal rapidity, yet with very satisfactory despatch. Westerly winds and clear weather prevail. Christian has seen two reindeer this week, and has shot a very few birds, and seven seals. As these creatures lie basking upon the ice, he crawls up to them behind a small calico screen, fitted upon a miniature sledge about a foot long, on which there is a rest for the muzzle of his rifle, and a slit in the calico through which he fires it. The seals afford an average weight of thirty pounds of excellent fresh meat, which we relish greatly, and consider much better suited to our present condition than such poor venison as reindeer would furnish at this season. A single hare has been shot; the white fur has nearly all disappeared, and left exposed the summer coat of dull lead colour. Several small birds not common to the northward are found here. Insects abound; the Doctor is perpetually in chase, unless busily occupied in grubbing up plants. Young is surveying the harbour. Hobson fully occupied with preparing the ship for sea. I have been giving some attention to the engines and boiler, and hope, with the help of the two stokers, to be able to make use of our steam power.

The men have received my hearty thanks

for their great exertions during the travelling period. I told them I considered every part of our search to have been fully and efficiently performed. Our labours have determined the exact position of the extreme northern promontory of the continent of America; I have affixed to it the name of Murchison, after the distinguished President of the Royal Geographical Society—the strenuous advocate for this "further search"—and the able champion of Lady Franklin when she needed all the support which private friendship and public spirit could bestow.

23*rd*.—The ice in Prince Regent's Inlet is broken up into pack, but the prevalence of easterly winds keeps it close in upon the shore. The ice about us is very much decayed, holes through it in many places. No reindeer seen this week, and only two seals procured; one of them shot by Christian, the other was killed by a bear, which ran off before Samuel could come within shot of him. A fox, a gull, a couple of ducks, and one or two lemmings, complete our game list for the week, yet our two Esquimaux are indefatigable in the pursuit. We eat all the birds and seals we can shoot, as well as mustard and cress as fast as we can grow it, but the quantity is very small. We sometimes refresh ourselves with a salad of sorrel-leaves, or roots of the little plant with lilac flower of snapdragon shape, named *Pedicularis hirsuta.*

The seine has been hauled in the narrow lake at the head of the harbour, but, as it was not well managed, only a dozen small trout were taken, though several were seen. We have tried for rock-cod, but without success. The relics of the lost expedition have been aired, exhibited to the crew, labelled, and packed away. The Doctor has been dredging lately. A record detailing our proceedings has been placed in a cairn upon the west point of Depôt Bay.

1*st August.*—A long continuance of unusually calm, bright, and warm weather has been favourable to our painting and cleaning the ship, scraping masts, and so forth. The result is that she looks unusually smart and gay, and our impatience to exhibit her, and *ourselves* at home is much increased. With the exception of a few gulls, and a duck, our hunters have shot nothing lately, although constantly out, either darting about in their kayaks or ranging over the hills; in fact there is nothing which they *can* shoot; the ducks are tolerably numerous, but extremely wild; the valleys are respectably clothed with vegetation, yet only one animal—a hare—has been seen. I was so fortunate as to shoot a snowy owl, the flesh of which was white and tender, but, to my palate, tasteless, although Petersen considers that "owl is the best beef in the country."

On Thursday night we found the harbour-ice to be quietly drifting out, of course taking us with it. The night was calm, the current in Bellot Strait very strong; we were almost helpless under the circumstances, and therefore felt the danger of our position. To warp the ship along the ice-edge, out of the way of the shore and rocks as it turned round and drifted along the cliffs to the westward, gave us some hours' occupation. At length it stuck fast between Fox Island and the main.

At turn of tide on Friday morning it began to drift eastward, and by this time being much broken up, and a breeze coming to our aid, we managed to extricate ourselves and reach a secure anchorage in Port Kenedy.

On Saturday night some ice that was left came drifting out of the inner harbour, and obliged us to slip our cable; but after a few hours we regained our berth in safety, and have since been undisturbed. There is no immediate prospect of escape, but we expect a prodigious smashing up of the ice whenever a strong wind springs up to set it in motion. To-day the steam was got up, and with the help of our two stokers I worked the engines for a short time. It is very cheering to know that we still have steam power at our command, although, by the deaths of poor Mr. Brand and Robert Scott, we were deprived of our engineer and engine-driver.

The mean temperature for July has been 40°·14, which is above the average for this region; the July temperatures have usually varied from 36° to 42°.

All are now in good health, but Hobson still a little lame. The issue of lemon-juice has been reduced to the ordinary allowance of half an ounce daily (as we have but little that is really good), lest another winter should become inevitable, which, I can devoutly say, may God forbid!

Monday night, 8th.—Very anxiously awaiting an opportunity to escape. We have constantly watched the ice from the neighbouring hills, including the lofty summit of Mount Walker—named after the Doctor, who was the first to ascend it (1123 feet)—from which Fury Point can be distinguished, but nothing very cheering has been seen. We had a N.E. gale, accompanied by rain and a considerable fall of the barometer, a few days ago; and as it blew freshly from the westward this morning, I went to a hill-top and saw that much ice had been broken up in Brentford Bay, and that there were streaks of water along the land between Possession Point and Hazard Inlet; this water, however, was not accessible to us.

The ice about Pemmican Rock was much in the same position as we found it last year, but Bellot Strait was perfectly clear. All the ice in this harbour, in Depôt Bay, and Hazard Inlet, is gone, by far the greater part having decayed, not drifted away.

Later in the day, and from loftier hill-tops, a good deal of water was seen off Cape Garry, and a water-sky beyond. It now blows very strongly from the S.W., the most desirable quarter; and as the anxious desire to escape has become oppressive, it is not to be wondered at that now our hopes have become extravagant. We may even make a start to-morrow! On the other hand, a careful examination of our provision store shows that, should we be obliged to spend another winter here, we must curtail our allowance of meat—fresh and salt—to three-quarters of a pound, and have to use but very indifferent lemon-juice. The spirits, I rejoice to say, will very shortly be entirely expended.

On the morning of the 3rd instant, when the rain ceased and N.E. gale sprang up, two claps of thunder were distinctly heard; this occurs but very rarely in these latitudes. There is ample occupation for the men, but not much for the officers; as for myself, I write a great deal, and work occasionally at our chart of discoveries; the only refreshment I indulge in is an occasional dive into packets of old letters. All yesterday the harbour was full of ice set in by southerly and westerly winds, and so closely packed that one might have walked over it to the shore; to-day it has nearly all drifted out again. The subjoined list will show what game we have been able to obtain by constant and arduous labour from the resources of these regions during nearly two years' sojourn.

Game List.

8 Mths. in the Pack, 1857-8.				11 Months in Port Kenedy, 1858-9.						
Bears.	Seals.	Dovekies.	Foxes.	Bears.	Deer.	Hares.	Foxes.	Ptarmigan.	Wild Fowl.	Seals.
2	73	38	1	2	8	9	19	82	98	18

At Port Kenedy several ermines and lemmings were also caught.
The ptarmigan all disappeared after 1st April.
Only 2 dovekies were seen, 1 in winter, and 1 in summer plumage.
A few seals were seen as early as the month of February.
Ducks, geese, and gulls were the usual kind of wild fowl killed.
During the 4 months occupied in sailing from Davis Strait to Bellot Strait, many looms and rotchies, and 5 or 6 bears were shot.

Wednesday, 10*th.*—The S.W. wind proved a good friend to us; by the morning of the 9th it had moved the ice off shore, and cleared away a passage for us out of Brentford Bay. We started under steam at eleven o'clock yesterday morning, and, passing round Long Island, made sail along the land towards Cape Garry, there being a channel about 2 or 3 miles wide between the pack and the shore.

The wind now failed us, and I experienced some little difficulty in the management of the engines and boiler; the latter primed so violently as to send the water over our top gallant yard, and the tail valve of the condenser by some means had got out of its seat, and admitted air to the condenser; but eventually we got the engines to work well, and steamed across Cresswell Bay during the night. The pack rested against Fury Point, and an east wind springing up, we made fast to a large grounded mass of ice in Adelaide Bay, about ¼ mile off shore, and in 3 fathoms' water, at eleven o'clock this morning. Having managed the engines for twenty-four consecutive hours, I was not sorry to get into bed. We were hardly out of Brentford Bay when fulmar petrels and white whales were seen; the first we have noticed for eleven and a half months. Dovekies are likewise abundant, and a seal has already been shot. Cresswell Bay is perfectly clear of ice, but this pale limestone land is the perfection of sterility, even with the rugged hills of Brentford Bay in lively recollection.

Upon the east side of Port Kenedy the bones of whales were found in two places a mile apart from each other; the lowest of them was 180 feet above the sea, the second was more than 300 feet high. The latter I examined, and found a jaw-bone, two ribs, a joint of the vertebræ, and fragments of other bones, all more or less buried in the soil, and much heavier than the bones of a recent animal; they lay within 40 or 50 yards of each other, and upon a little flat patch of rather rich earth, a rocky hill above, and steep slope below;—they are also nearly a mile inland.

Of the traces which we have left behind us, the most considerable are the graves of our two shipmates within the western point of our little harbour; they were tastefully sodded round, and planted over with the usual Arctic flowers. There is our record in a conspicuous cairn at the west point of Depôt or Transition Bay: we left also three cases of pemmican near the east end of the Long Lake, and our travelling boat near its west end, at the head of False Strait.

Monday, 15*th*.—Strong east winds, with much rain, have imprisoned us here for the last four days, and driven the whole pack close in, completely filling up Cresswell Bay. We remain fast to the grounded ice, which shields us from pressure, otherwise we should have been driven irretrievably on shore. A couple more seals and a white whale have been shot; the latter measured 13½ feet long, and proved to be a female of ordinary dimensions, and of an uniform cream colour; the eyes are extremely small, and orifices of the ears scarcely large enough to admit a crow-quill. We dined off steaks of the flesh, and prefer it to seal, which it very much resembles, but is not quite so tender; the skin is greatly prized by the Greenlanders as an antiscorbutic; it is a sort of gristly gelatinous substance, nearly half an inch thick, and possessing very little taste; fried and eaten with fish-sauce, it reminded me of cod sound, though not so good.

The blubber fills two twenty-gallon casks; it produces oil of a quality superior to seal oil; not an ounce of the flesh or skin of this huge animal has been thrown away, the men having a wholesome dread of scurvy, and unbounded confidence in "blood-meat," such as this! The Doctor has picked up a few fossils very similar to those formerly brought home from Port Leopold.

To our great joy the east wind died away this morning, and immediately a west wind sprang up, which very quickly freshened to a smart gale. At four o'clock this afternoon we were able to make sail, the ice having moved about 3 miles off shore. Passed within a mile of Fury Beach two hours afterwards, and saw the framing of the house, the boats and casks very distinctly.

17*th.*—After passing Fury Beach it fell calm, so we steamed up as far as Batty Bay. On Tuesday afternoon we were off Port Leopold, running fast, when thick fog came on, and we got involved in loose ice, and seriously damaged our rudder. The boats and stores at Port Leopold appeared to remain as we left them last year. The flag-staff on the summit of North-east Cape (over Whale Point) is still standing, but not erect.

Fog and ice obstructed our progress during the night; but this morning when I came on deck at eight o'clock, the day was bright, clear, and charming; no ice visible, except about Leopold Island, which was now some miles behind us. Towards evening the wind became contrary.

Sunday evening, 21*st*.—At sea—out of sight of land!

On the 19th we were somewhat delayed by loose ice off Cape Hay, but by noon yesterday were close off Cape Burney, and whilst almost becalmed there, a mother bear swam off to us with two interesting cubs about the size of very large dogs—foolish creatures! a volley of rifles decided their fate in a very few seconds. Not finding any whaling vessels off Pond's Inlet, the land-ice which shelters the whales having all disappeared, we therefore concluded that the whalers had left in consequence, so, without seeking for them further south, at once changed our course for Disco.

To-day only a few icebergs have been seen. There is a good deal of swell, so we tumble about. Roast *veal* has appeared amongst the delicacies of our table since the battue of yesterday, and Christian has asked for a portion of the old bear to carry home to his mother. Bear's flesh is really considered a delicacy in Greenland.

25*th*.—Becalmed off Hare Island, and getting the steam ready. We are only 108 miles from Godhavn, and the anxiety to clutch our letters has become intolerable. No pack-ice has been met with in our passage across Baffin's Bay, but many icebergs. This morning the lofty snow-clad land of Noursoak and Disco was beautifully distinct; and at the same time the wind died away, leaving us, at least, the opportunity to contemplate at our *leisure*, their gloomy grandeur.

26*th*.—Steamed for ten hours last night. Fair winds and calms have alternated since then, but this evening we are within 20 miles, and hope soon to get into port. I have been reading over Young's report of his spring journey. It comprises seventy-eight days of sledge-travelling, and certainly under most discouraging circumstances. Leaving the ship on 7th April, he crossed the western strait to Prince of Wales' Land, and thence traced its shore to the south and west. On reaching its southern termination—Cape Swinburne, so named in honour of Rear-Admiral Swinburne, a much-esteemed friend of Sir J. Franklin, and one of the earliest supporters of this final expedition—he describes the land as extremely low, and deeply covered with snow, the heavy grounded hummocks which fringed its monotonous coast alone indicating the line of demarcation betwixt land and sea. To the north-east of this terminal cape the sea was covered with level floe formed in the fall of last year, whilst all to the north-westward of the same cape was pack consisting of heavy ice-masses, formed perhaps years ago in far distant and wider seas.

Young attempted to cross the channel which he discovered between Prince of Wales' Island and Victoria Land; but from the rugged nature of the ice, found it quite impracticable with the means and time remaining at his disposal. Young expresses his firm conviction that this channel is so constantly choked up with unusually heavy ice as to be quite unnavigable; it is, in fact, a *continuous ice-stream* from the N.W. His opinion coincides with my own, and with those of Captains Ommanney and Osborn, when those officers explored the north-western shores of Prince of Wales' Land in 1851.

Fearing that his provisions might run short, he sent back one sledge with four men, and continued his march with only one man and the dogs for forty days! They were obliged to build a snow-hut each night to sleep in, as the tent was sent back with the men; but latterly, when the weather became more mild, they preferred sleeping on the sledge, as the constructing of a snow-hut usually occupied them for two hours. Young completed the exploration of this coast beyond the point marked upon the charts as Osborn's farthest, up nearly to lat. 73° N., but no cairn was found. Young, however, recognised the remarkably shaped conical hills spoken of by Osborn, when he at his farthest, in 1851, struck off to the westward.

The coast-line throughout was extremely low; and in the thick disagreeable weather which he almost constantly experienced, it was often a matter of great difficulty to prevent straying off the coast-line inland. He commenced his return on 11th May, and reached the ship on 7th June, in wretched health and depressed in spirits.

Directly his health was partially re-established, he, in spite of the Doctor's remonstrances, as I have before said, again set out on the 10th with his party of men and the dogs, to complete the exploration of both shores of the continuation of Peel Sound, between the position of the 'Fox' and the points reached by Sir James Ross in 1849, and Lieutenant Browne in 1851. This he accomplished without finding any trace of the lost expedition, and the parties were again on board by 28th June. The ice travelled over in this last journey was almost all formed last autumn.

The extent of coast-line explored by Captain Young amounts to 380 miles, whilst that discovered by Hobson and myself amounts to nearly 420 miles, making a total of 800 geographical miles of new coast-line which we have laid down.

Hobson's report is a minute record of all that occurred during his journey of seventy-four days, and includes a list of all the relics brought on board, or seen by him. He suffered very severely in health: when only ten days out from the ship, traces of scurvy appeared; when a month absent he walked lame; towards the latter end of the journey he was compelled to allow himself to be dragged upon the sledge, not being able to walk more than a few yards at a time; and on arriving at the ship on the 14th June, poor Hobson was unable to stand. How strongly this bears upon the last sad march of the lost crews! And yet Hobson's food throughout the whole journey was pemmican of the very best quality, the most nutritious description of food that we know of, and varied occasionally by such game as they were able to shoot. In spite of this fresh-meat diet, scurvy advanced with rapid strides.

After leaving me at Cape Victoria, he says—"No difficulty was experienced in crossing James Ross Strait. The ice appeared to be of but one year's growth; and although it was in many places much crushed up, we easily found smooth leads through the lines of hummocks;

many very heavy masses of ice, evidently of foreign formation, have been here arrested in their drift: so large are they that, in the gloomy weather we experienced, they were often taken for islands."

Again, at Cape Felix, he observes,—"The pressure of the ice is severe, but the ice itself is not remarkably heavy in character; the shoalness of the coast keeps the line of pressure at a considerable distance from the beach: to the northward of the island the ice, as far as I could see, was very rough, and crushed up into large masses." Here we notice the gradual change in the character of the ice as Hobson left the Boothian shore and advanced towards Victoria Strait. The "very heavy masses of ice, evidently of foreign formation," had drifted in from the N.W. through M'Clure Strait; Victoria Strait was full of it; and Hobson's description of the ice he passed over clearly illustrates how Franklin, leaving clear water behind him, pressed his ships into the pack when he attempted to force through Victoria Strait. How very different the result *might* and probably *would* have been had he known of the existence of a ship-channel, sheltered by King William Island from this tremendous "polar pack"!

Hobson left King William Island on the last

day of May, having spent thirty-one days on its desolate shores. During that period one bear and five willow grouse were shot; one wolf and a few foxes were seen. One poor fox was either so desperately hungry, or so charmed with the rare sight of animated beings, that he played about the party until the dogs snapped him up, although in harness and dragging the sledge at the time. A few gulls were seen, but not until after the first week in June.

I have already explained how Hobson found the records and the boat: he exercised his discretionary power with sound judgment, and completed his search so well, that, in coming over the same ground after him, I could not discover any trace that had escaped him.

I quite agree with him that there may be many small articles beneath the snow; but that cairns, graves, or any conspicuous objects could exist upon so low and uniform a shore, without our having seen them, is *almost* impossible.

Sunday evening, 29*th*.—Calm, warm, lovely weather; and we are thoroughly enjoying it in the quiet security of Lievely harbour, or Godhavn. Although Friday night was dark, we managed to find out the harbour's mouth, and slowly steamed into it. The inhabitants were awoke by Petersen demanding our letters, but great indeed was our disappointment at finding only a very few letters and two or three papers, and these for the officers only! It appears that on the arrival of the whalers in early spring, the ice prevented their usual communication with the settlement, therefore the letters on board of them were unavoidably carried northward. Some few, however, which came out in the 'Truelove,' were landed at the neighbouring settlement of Noursoak, and from thence were sent back to Godhavn.

It is rather a nervous thing opening the first letters after a lapse of more than two years. We received them in our beds at three o'clock in the morning; and when we met at breakfast were able, thank God! to congratulate each other upon the receipt of cheering home news. Lady Franklin and Miss Cracroft wrote to me from Bournemouth in March last. They have travelled more than we have, I think, having visited almost all the countries bordering the Mediterranean and Black Seas, posted through the Crimea, and steamed up the Danube! I am much gratified to learn that I have been elected a member of the Royal Yacht Squadron during my absence.

Yesterday morning I called upon the inspector, Mr. Olrik, who has been home to Denmark since I saw him last spring. In the autumn he took Mrs. Olrik and his family to Copenhagen, and has but just returned alone. He received me with his usual kindness, and promised me such supplies as we require. It so happens that none of my expected business letters have arrived, so that I am not accredited in the slightest degree, nor is there any hint thrown out as to where I am to take the 'Fox.' Mr. Olrik gave me a large bundle of the 'Illustrated London News,' which was exceedingly acceptable, and told us that Austria was at war with France and Sardinia. By the latest news a battle had been fought and won by the latter Powers. Most fortunately a 'Navy List' had come out to Hobson, otherwise I think we should have been utterly brokenhearted. We study its pages daily, and delight in noticing the advancement of our many friends.

1*st Sept.*, *Thursday night.*—At sea, on *the passage*, and already enjoying, by anticipation, the pleasures of home! Five busy days were spent in Godhavn, supplying our little wants, in as far as they could be supplied, including 100 gallons of light beer. The natives were very useful, the men bringing off water, stone ballast, and sand, and a troop of Esquimaux girls scrubbing the paintwork and the decks.

Each evening the men went on shore, taking with them a very limited quantity of rum-punch for the ladies, and danced for several hours in a large store; whilst the officers and myself spent the time with Mr. Olrik or the other Danish gentlemen—Messrs. Andersen, Bulbrue, and Tyner. Nothing could exceed their kindness to us, whilst their good humour and their anecdotes, sometimes expressed in quaint English, greatly amused us. We shall always retain very agreeable recollections of Godhavn; twice has it been to us an Arctic home.

Mr. Petersen's nieces, the belles of the place, came on board (Miss Sophia with scented cambric handkerchief and gloves—in other respects she adheres to the Esquimaux costume); they were pleased with the organ, although it is rather out of repair, and they sang together very sweetly for us. Our Esquimaux shipmates, Christian and Samuel, were discharged, and, by their own request, their wages given in charge to Mr. Olrik and Mr. Bulbrue; they seemed to understand the importance of husbanding their wealth. Christian said he thought it would not be all spent under three years. First of all he intended buying a rifle for his brother, and then some wood to build a house for himself.

I was gratified very much when I heard them say that the men had treated them very well—"all the same as brothers;" and they really seemed sorry to leave the ship; they would come on board and look gravely about at everything as if regretting the coming separation. Even our poor dogs seemed to think the ship their natural abode; although landed at the settlement, they soon ran round the harbour to the point nearest to the ship, and there, upon the rocks, spent the whole period of our stay.

On Tuesday night we set off some fireworks on shore to amuse the natives, for I intended sailing next day, but the wind prevented my doing so. The last day was spent in the interchange of presents between our Danish friends and ourselves; indeed, the sincere hearty good feeling which existed between every individual in the 'Fox' and the inhabitants of the settlement was as gratifying as apparent. Almost the only fresh supplies obtained here were rock cod and salmon-trout from Disco fiord. During our stay the weather was delightful; indeed, it was the first really fine weather they had experienced at Godhavn during the present season, the summer having been cold and wet.

10*th Sept., Saturday night.*—To-day we passed to the eastward of Cape Farewell, but about 100 miles to the south of it. The last iceberg was seen to-day; and now we are running along swiftly before a pleasant N.W. breeze. Hitherto we have had every variety of wind and weather, from a calm to a gale, but generally the wind has been favourable. The change of temperature is already very perceptible.

Saturday night, 17th Sept.—A week of favourable gales has brought us from Cape Farewell to within 400 miles of the Land's End, or about 1100 miles of distance. But such rough weather is not pleasant in so small a vessel, however much "like a duck" she may be; and our two years' sojourn in the still waters of the frozen North has made us very susceptible of the change.

CONCLUSION.

We sailed all the way home from Greenland, yet the 'Fox' made the passage in only nineteen days, arriving in the English Channel on 20th September; on the evening of the 21st I reached London (having landed at Portsmouth), and made known to the Admiralty the result of my voyage.

On the 23rd September the 'Fox' was taken into dock at Blackwall; and, through the kindness and promptitude of the Lords of the Admiralty, I was enabled on the 27th, when the crew were assembled for the last time, to present the Arctic medal to such of my companions as had not already received it for previous Arctic service, and also to inform Lieutenant Hobson that his promotion to the rank of Commander would speedily take place.

I will not intrude upon the reader, who has followed me through the pages of this simple narrative, any description of my feelings on finding the enthusiasm with which we were all received on landing upon our native shores.

The blessing of Providence had attended our efforts, and more than a full measure of approval from our friends and countrymen has been our reward. For myself the testimonial given me by the officers and crew of the 'Fox' has touched me perhaps more than all. The purchase of a gold chronometer, for presentation to me, was the first use the men made of their earnings; and as long as I live it will remind me of that perfect harmony, that mutual esteem and goodwill, which made our ship's company a happy little community, and contributed materially to the success of the expedition.

The names I have given to my discoveries are, with the exception of those by which I have endeavoured to honour the members of the lost expedition, the names of active supporters of the recent search, and friends of Franklin and his companions, though such names are far from exhausting the number of those who have the highest claims to distinction on both grounds.

It will be observed that I have refrained from repeating names which have already been commemorated by preceding commanders, and which therefore are already in our charts. Besides the individuals already mentioned in the narrative, Sir Thomas D. Acland, one of the most zealous promoters of the search, both in and out of the House of Commons; Monsieur De la Roquette, Vice-President of the Geographical Society of Paris, and author of an interesting biography of Franklin; Rear-Admiral Fitzroy; and Major-General Pasley, R.E., stand high amongst those whom it has been my privilege to honour.

Although much talent has been brought to bear upon the deciphering of the letters found in a pocketbook near Cape Herschel (page 274 *ante*), yet, from their being so very much defaced by time, only a few detached sentences have been made out, and these do not in the slightest degree refer to the proceedings of the lost expedition.

It will be seen that I have noticed (page 288) the discrepancy between the number of souls accounted for by the Point Victory Record, and the generally received opinion that 138 individuals sailed in the 'Erebus' and 'Terror.'

I am now enabled to state, on the authority of the Admiralty, that only one hundred and thirty-four individuals left the United Kingdom, and of these five men subsequently returned: one by H.M.S. 'Rattler,' and four by the transport 'Barretto Junior;' so that only one hundred and twenty-nine—the exact number mentioned in the record—actually entered the ice. The five invalids were—

From H.M.S. 'Terror,' John Brown, Able Seaman.
" Robert Carr, Armourer.
" James Elliot, Sailmaker.
" William Aitken, Marine.
From H.M.S. 'Erebus,' Thomas Birt, Armourer.

The relics we have brought home have been deposited by the Admiralty in the United Service Institution, and now form a national memento—the most simple and most touching—of those heroic men who perished in the path of duty, but not until they had achieved the grand object of their voyage,—the *Discovery of the North-West Passage.*

London, 24*th Nov*. 1859.

APPENDIX.

No. I.

A LETTER TO VISCOUNT PALMERSTON, K.G., &c., FROM LADY FRANKLIN.

MY LORD, 60, Pall Mall, December 2, 1856.

I trust I may be permitted, as the widow of Sir John Franklin, to draw the attention of Her Majesty's Government to the unsettled state of a question which a few months ago was under their consideration, and to express a well-grounded hope that a final effort may be made to ascertain the fate and recover the remains of my husband's expedition.

Your Lordship will allow me to remind you that a Memorial* with this object in view (of which I enclose a printed copy) was early in June last presented to, and kindly received by you. It had been signed within forty-eight hours by all the leading men of science then in London who had an opportunity of seeing it, and might have received an indefinite augmentation of worthy names had not the urgency of the question forbidden delay. To the above names were appended those of the Arctic officers who had been personally engaged in the search, and who, though absent, were known to be favourable to another effort for its completion. And though that united application obtained no immediate result, it was felt, and by no one more strongly than myself, that it never could be utterly wasted.

---

* See Appendix II.

I venture also to allude to a letter of my own addressed to the Lords Commissioners of the Admiralty in April last, and a copy of which accompanied, I believe, the Memorial to your Lordship, wherein I earnestly deprecated any premature adjudication of the reward claimed by Dr. Rae, on the ground that the fate of my husband's expedition was not yet ascertained, and that it was due both to the living and the dead to complete a search which had been hitherto pursued under the greatest disadvantage, for want of the clue which was now for the first time in our hands.

The Memorial above alluded to, and my own letter of earlier date, had not yet received any reply, when, in the month of July, the Lords of the Admiralty caused prompt inquiries to be made as to the possibility of equipping a ship at that advanced season, in time for effective operations in the field of search. The result was that it was pronounced to be too late, and the subject was dismissed for that season.

Upon this I addressed a letter to the Board (of which I take the liberty to enclose a copy), respectfully showing that by this unfortunate delay the opportunity had also been taken from me of sending out a vessel at my own cost, a measure which I had previously felt myself obliged to state to their Lordships would be the alternative of any adverse decision on their part. I pleaded therefore, as the only remedy for the loss of an entire summer season, that the route by Behring Straits was by some of the most competent Arctic officers considered preferable to the eastern route, and that the equipment of a vessel for this direction need not take place before the close of the year.

In reply, their Lordships caused me to be informed that "they had come to the decision not to send any expedition to the Arctic regions in the present year."

This communication, however, was in answer merely to my own letter. The Memorialists had as yet received no reply, and accordingly the President of the Royal Society put a question respecting the Memorial in the House of Lords at the close of the session, which drew from one of Her Majesty's Ministers (Lord Stanley), after some preliminary observations, the assurance that Her Majesty's Government would give the subject their serious consideration during the recess. I may be permitted to add, that, in the conversation which followed, Lord Stanley expressed himself as very favourably disposed towards a proposition made to him by Lord Wrottesley, that, in the event of there being no Government expedition, I should be assisted in fitting out my own expedition; an assurance which Lord Wrottesley had the kindness to communicate to me by letter.

But, my Lord, as nothing has occurred within the last few months to weaken the reasons which induced the Admiralty, early in July last, to contemplate another final effort, and as they put it aside at that time on the sole ground that it was too late to equip a vessel for that season, I trust it will be felt that I am not endeavouring to re-open a closed question, but merely to obtain the settlement of one which has not ceased to be, and is even now, under favourable consideration. The time has arrived, however, when I trust I may be pardoned for pressing your Lordship, with whom I believe the question rests, for a decision, since by further delay even my own efforts may be paralysed.

I have cherished the hope, in common with others, that we are not waiting in vain. Should, however, that decision unfortunately throw upon me the responsibility and the cost of sending out a vessel myself, I beg to assure your Lordship that I shall not shrink, either from that weighty responsibility, or from the sacrifice of my entire available fortune for the purpose, supported as I am in my convictions by such high authorities as those whose opinions are on record in your Lordship's hands, and by the hearty sympathy of many more.

But before I take upon myself so heavy an obligation, it is my bounden duty to entreat Her Majesty's Government not to disregard the arguments which have led so many competent and honourable men to feel that our country's honour is not satisfied, whilst a mystery which has excited the sympathy of the civilised world remains uncleared. Nor less would I entreat you to consider what must be the unsatisfactory consequences, if any endeavours should be made to quench all further efforts for this object.

It cannot be that this long-vexed question would thereby be set at rest, for it would still be true that in a certain circumscribed area within the Arctic circle, approachable alike from the east and from the west, and sure to be attained by a combination of both movements, lies the solution of our unhappy countrymen's fate. While such is the case, the question will never die. I believe that again and again would efforts be made to reach that spot, and that the Government could not look on as unconcerned spectators, nor be relieved in public opinion of the responsibility they had prematurely cast off.

But I refrain from pursuing this argument, though, if any illustration were wanting of its truth, I think it might be found in the events that are passing before our eyes.

It is now about two years ago that one of Her Majesty's Arctic ships was abandoned in the ice. In due time this ship floated away, was picked up by an American whaler, carried into an American port, and (all property in her having been relinquished by the Admiralty) was purchased of her rescuers by the American Government, by whom she has been lavishly re-equipped, and is now on her passage to England, a free gift to the Queen. The 'Resolute' is about to be delivered up in Portsmouth harbour, not merely in evidence of the cordial relation existing between the two countries, but as a lively token of the deep interest and sympathy of the Americans in that great cause of humanity in which they have so nobly borne their part. The resolution of Congress expressly states this motive, and indeed there could be no other, as it is well known that for any purpose but the Arctic service those expensive equipments would be perfectly useless and require removal.

My Lord, you will not let this rescued and restored ship, emblematic of so many enlightened and generous sentiments, fail, even partially, in her significant mission. I venture to hope that she will be accepted in the spirit in which she is sent. I humbly trust that the American people, and especially that philanthropic citizen who has spent so largely of his private fortune in the search for the lost ships, and to whom was committed by his Government the entire charge of the equipment of the 'Resolute,' will be rewarded for this signal act of sympathy, by seeing her restored to her original vocation, so that she may bring back from the Arctic seas, if not some living remnant of our long-lost countrymen, yet at least the *proofs* that they have nobly perished.

I need not add that we have as yet no proofs, whatever may be our melancholy forebodings. That such is the fact, in a legal point of view, is shown by a case now or lately pending in the Scotch Courts, in which the right of succession to a considerable property is not admitted, on account of the absence of all but conjectural testimony. In this aspect of the question I have no personal interest, but it is one that may not be deemed unworthy of your Lordship's attention, combined as it must be with the fact that our most experienced Arctic officers are willing to stake their reputation upon the feasibility of reaching the spot where so many secrets lie buried, if only they are supplied with the adequate means.

It would be a waste of words to attempt to refute again the main objections that have been urged against a renewed search, as involving extraordinary danger and risking life. The safe return of our officers and men cannot be denied, neither will it be disputed that each succeeding year diminishes the risk of casualty; and indeed, I feel it would be especially superfluous and unseasonable to argue against this particular objection, or against the financial one which generally accompanies it, at a moment when new expeditions for the glorious interests of science, and which every true lover of science and of his country must rejoice in, are contemplated for the interior of Africa and other parts which are far less favourable to human life than the icy regions of the north.

But with respect to expenditure, I may perhaps be allowed, as I have alluded to that topic, again to call to your Lordship's attention that the 'Resolute' is ready equipped for Arctic service by the munificence of another nation, and to add that other Arctic ships, equally well fitted for the purpose, are lying useless in Her Majesty's dockyards, along with accumulated Arctic stores brought back by the late expeditions, and therefore long since included in the navy estimates; and which, besides, are available only for Arctic service, and, if sold, would be bought at only nominal prices. In addition to the above sources of supply are those already existing on the Arctic shores, which are now studded with depôts of provisions and fuel, left from the last and former expeditions, and fit as ever for use, because of the conservative properties of the climate.

But even were the expenditure greater than can thus reasonably be expected, I submit to your Lordship that this is a case of no ordinary exigency. These 135 men of the 'Erebus' and 'Terror' (or perhaps I should rather say the greater part of them, since we do not yet know that there are no survivors) have laid down their lives, after sufferings doubtless of unexampled severity, in the service of their country, as truly as if they had perished by the rifle, the cannon-ball, or the bayonet. Nay more,—by attaining the northern and already-surveyed coast of America, it is clear that they solved the problem which was the object of their labours, or, in the beautiful words of Sir John Richardson, that "they forged the last link of the North-West passage with their lives."

Surely, then, I may plead for such men, that a careful search be made for any possible survivor, that the bones of the dead be sought for and gathered together, that their buried records be unearthed, or recovered from the hands of the Esquimaux, and above all, that their last written words, so precious to their bereaved families and friends, be saved from destruction. A mission so sacred is worthy of a Government which has grudged and spared nothing for its heroic soldiers and sailors in other fields of warfare, and will surely be approved by our gracious Queen, who overlooks none of Her loyal subjects suffering and dying for their country's honour.

This final and exhausting search is all I seek in behalf of the first and only martyrs to Arctic discovery in modern times, and it is all I ever intend to ask.

But if, notwithstanding all I have presumed to urge, Her Majesty's Government decline to complete the work they have carried on up to this critical moment, but leave it to private hands to finish, I must then respectfully request that measure of assistance in behalf of my own expedition which I have been led to expect on the authority of Lord Stanley, as communicated to me by Lord Wrottesley, and on that of the First Lord of the Admiralty, as communicated to Colonel Phipps in a letter in my possession.

It is with no desire to avert from myself the sacrifice of my own funds, which I devote without reserve to the object in view, that I plead for a liberal interpretation of those communications, but I owe it to the conscientious and high-minded Arctic officers who have generously offered me their services, that my expedition should be made as efficient as possible, however restricted it may be in extent. The Admiralty, I feel sure, will not deny me what may be necessary for this purpose, since, if I do all I can with my own means, any deficiencies and shortcomings of a private expedition cannot I think be justly laid to my charge.

In conclusion, I would earnestly entreat of Her Majesty's Government, while this subject is still under deliberation, that they would be pleased to obtain the opinions of those persons who, in consequence of their practical knowledge and vast experience, may be considered best qualified to express them in the present emergency. And as it must be in the ranks of those officers who would naturally be selected for command of any final expedition that these qualifications will most assuredly be found, I trust I may be pardoned for directing your Lordship's attention to the names (which I put down in the order of their seniority) of Captains Collinson, Richards, McClintock, Maguire, and Osborn. All these officers have passed winter after winter in Arctic service, have carried out those skilful sledge operations which have added so much to our knowledge of Arctic Geography, and have ever, in the exercise of combined courage and discretion, avoided disaster, and brought home their crews in health and safety.

I commit the prayer of this letter, for the length of which I beg much to apologize, to your Lordship's patient and kind consideration, feeling assured that, however the burden of it may pall upon the ear of some, who apparently judge of it neither by the heart nor by the head, you will not on that, or on any light ground, hastily dismiss it. Rather may you be impelled to feel that the shortest and surest way to set the importunate question at rest, is to submit it to that final investigation which will satisfy the yearnings of surviving relatives and friends, and, what is justly of higher import to your Lordship, the credit and honour of the country.

I have the honour to be, &c.,

JANE FRANKLIN.

The Right Hon. Viscount Palmerston, K.G.

No. II.

MEMORIAL TO THE RIGHT HON. VISCOUNT PALMERSTON, M.P., G.C.B.

London, June 5th, 1856.

IMPRESSED with the belief that Her Majesty's missing ships, the 'Erebus' and 'Terror,' or their remains, are still frozen up at no great distance from the spot whence certain relics of Sir John Franklin and his crews were obtained by Dr. Rae,—we whose names are undersigned, whether men of science and others who have taken a deep interest in Arctic discovery, or explorers who have been employed in the search for our lost countrymen, beg earnestly to impress upon your Lordship the desirableness of sending out an Expedition to satisfy the honour of our country, and clear up a mystery which has excited the sympathy of the civilised world.

This request is supported by many persons well versed in Arctic surveys, who, seeing that the proposed Expedition is to be directed *to one limited area only*, are of opinion that the object is attainable, and with little risk.

We can scarcely believe that the British Government, which to its great credit has made so many efforts in various directions to discover even the route pursued by Franklin, should cease to prosecute research, now that the locality has been clearly indicated where the vessels or their remains must lie,—including, as we hope, records which will throw fresh light on Arctic geography, and dispel the obscurity in which the voyage and fate of our countrymen are still involved.

Although most persons have arrived at the conclusion that there can now be no survivors of Franklin's Expedition, yet there are eminent men in our own country and in America who hold a contrary opinion. Dr. Kane, of the United States, for example, who has distinguished himself by pushing farther to the north in search of Franklin than any other individual, and to whom the Royal Geographical Society has recently awarded its Founders' Gold Medal, thus speaks (in a letter to the benevolent Mr. Grinnell):—"I am really in doubt as to the preservation of human life. I well know how glad I would have been, had my duty to others permitted me, to have taken refuge among the Esquimaux of Smith Strait and Etah Bay. Strange as it may seem to you, we regarded the coarse life of these people with eyes of envy, and did not doubt but that we could have lived in comfort upon their resources. It required all my powers, moral and physical, to prevent my men from deserting to the Walrus Settlements, and it was my final intention to have taken to Esquimaux life had Providence not carried us through in our hazardous escape."

But passing from speculation, and confining ourselves alone to the question of finding the missing ships or their records, we would observe that no land Expedition down the Back River, like that which, with great difficulty, recently reached Montreal Island, can satisfactorily accomplish the end we have in view. The frail birch-bark canoes in which Mr. Anderson conducted his search with so much ability, the dangers of the river, the sterile nature of the tract near its embouchure, and the necessary failure of provisions, prevented the commencement, even, of such a search as can alone be satisfactorily and thoroughly accomplished by the crew of a man-of-war,—to say nothing of the moral influence of a strong armed party remaining in the vicinity of the spot until the confidence of the natives be obtained.

Many Arctic explorers, independent of those whose names are appended, and who are absent on service, have expressed their belief that there are several routes by which a *screw*-vessel could so closely approach the area in question as to clear up all doubt.

In respect to one of these courses, or that by Behring Strait, along the coast of North America, we know that a single sailing vessel passed to Cambridge Bay, within 150 miles of the mouth of the Back River, and returned home unscathed,—its commander having expressed his conviction that the passage in question is so constantly open that ships can navigate it without difficulty in one season. Other routes, whether by Regent Inlet, Peel Sound, or across from Repulse Bay, are preferred by officers whose experience in Arctic matters entitles them to every consideration; whilst in reference to two of these routes it is right to state that vast quantities of provisions have been left in their vicinity.

Without venturing to suggest which of these plans should be adopted, we earnestly beg your Lordship to sanction without delay such an expedition as, in the judgment of a Committee of Arctic Voyagers and Geographers, may be considered best adapted to secure the object.

We would ask your Lordship to reflect upon the great difference between a clearly-defined voyage to a narrow and circumscribed area, within which the missing vessels or their remains must lie, and those formerly necessarily tentative explorations in various directions, the frequent allusions to the difficulty of which, in regions far to the north of the voyage now contemplated, have led persons unacquainted with geography to suppose that such a modified and limited attempt as that which we propose involves farther risk and may call for future researches. The very nature of the former expeditions exposed them, it is true, to risk, since regions had to be traversed which were totally unknown; while the search we ask for is to be directed to a circumscribed area, the confines of which have already been reached without difficulty by one of Her Majesty's vessels.

Now, inasmuch as France, after repeated fruitless

efforts to ascertain the fate of La Perouse, no sooner heard of the discovery of some relics of that eminent navigator, than she sent out a Searching Expedition to collect every fragment pertaining to his vessels, so we trust that those Arctic researches which have reflected much honour upon our country may not be abandoned at the very moment when an explanation of the wanderings and fate of our lost navigators seems to be within our grasp.

In conclusion, we further earnestly pray that it may not be left to the efforts of individuals of another and kindred nation already so distinguished in this cause, nor yet to the noble-minded widow of our lamented friend, to make an endeavour which can be so much more effectively carried out by the British Government.

We have the honour to be, &c.,

F. Beaufort,
R. I. Murchison,
F. W. Beechey,
Wrottesley,
E. Sabine,
Egerton Ellesmere,
W. Whewell,
R. Collinson,
W. H. Sykes,
C. Daubeny,
J. Fergus,
P. E. de Stzrelecki,
W. H. Smyth,
A. Majendie,
R. Fitzroy,
E. Gardiner Fishbourne,
R. Brown,
G. Macartney,
L. Horner,
W. H. Fitton,
Lyon Playfair,
T. Thorp,
C. Wheatstone,
W. J. Hooker,
J. D. Hooker,
J. Arrowsmith,
P. La Trobe,
W. A. B. Hamilton,
R. Stephenson,
J. E. Portlock,
C. Piazzi Smyth,
C. W. Pasley,
G. Rennie,
J. P. Gassiot,
G. B. Airy,
J. F. Burgoyne.

The following officers of the Royal Navy, who have been employed in the search after Franklin, and who are now absent from London, have previously expressed themselves to be favourable to the final expedition above recommended:—

Captains Sir JAMES C. ROSS, and Sir EDWARD BELCHER;
Commodore KELLETT;
Captains AUSTIN,
BIRD,
OMMANNEY,
Sir ROBERT M'CLURE,
SHERARD OSBORN,
INGLEFIELD,
Captains MAGUIRE,
M'CLINTOCK, and
RICHARDS;
Commanders ALDRICH,
MECHAM,
TROLLOPE, and
CRESSWELL;
Lieutenants HAMILTON and PIM.

No. III.

LIST OF RELICS OF THE FRANKLIN EXPEDITION

Brought to England in the 'Fox' by Captain M'Clintock.

Relics brought from the boat found in lat. 69° 08′ 43″ N., long. 99° 24′ 42″ W., upon the West Coast of King William Island, May 30, 1859:—

Two double-barrelled guns, one barrel in each is loaded. Found standing up against the side in the after part of the boat.

A small Prayer Book; cover of a small book of 'Family Prayers;' 'Christian Melodies,' an inscription within the cover to "G. G." (Graham Gore?); 'Vicar of Wakefield;' a small Bible, interlined in many places, and with numerous references written in the margin; a New Testament in the French language.

Two table knives with white handles—one is marked "W. R.;" a gimlet; an awl; two iron stanchions, 9 inches long, for supporting a weather cloth, which was round the boat.

26 pieces of silver plate—11 spoons, 11 forks, and 4 teaspoons; 3 pieces of thin elmboard (tingles) for repairing the boat, and measuring 11 inches by 6 inches, and 3-10ths inch thick.

Piece of canvas:—Bristles for shoemaker's use, bullets, short clay pipe, roll of waxed twine, a wooden button, small piece of a port-fire, two charges of shot tied up in the finger of a kid glove, fragment of a seaman's blue serge frock. Covers of a small Testament and Prayer Book, part of a grass cigar-case, fragment of a silk handkerchief, thread-case, piece of scented soap, three shot charges in kid glove fingers, a belted bullet, a piece of silk pocket handkerchief. Two pairs of goggles, made of stout leather and wire gauze, instead of glass; a sailmaker's palm, two small brass pocket compasses, a snooding line rolled up on a piece of leather, a needle and thread case, a bayonet scabbard altered into a sheath for a knife, tin water bottle for the pocket, two shot pouches (full of shot).

Three spring hooks of sword belts, a gold lace band, a piece of thin gold twist or cord, a pair of leather goggles with crape instead of glass; a small green crape veil.

Two small packets of blank cartridge in green paper, part of a cherry-stick pipe stem, piece of a port-fire, a few copper nails, a leather bootlace, a seaman's clasp-knife, two small glass stoppered bottles (full), three glasses of spectacles, part of a broken pair of silver spectacles, German silver pencil-case, a pair of silver (?) forceps, such as a naturalist might use for holding or seizing small insects, &c.; a small pair of scissors rolled up in blank paper, and to which adheres a printed Government paper, such as an officer's warrant or appointment; a spring hook of a sword belt, a brass charger for holding two charges of shot.

A small bead purse, piece of red sealing-wax, stopper of a pocket flask, German silver top and ring, brass matchbox, one of the glasses of a telescope, a small tin cylinder, probably made to hold lucifer matches; a linen bag of percussion caps of three sizes, a very large and old-fashioned kind, stamped "Smith's patent;" a cap with a flange similar to the present musket caps used by Government, but smaller; and ordinary sporting caps of the smallest size.

Five watches.

A pair of blue glass spectacles, or goggles, with steel frame, and wire gauze encircling the glasses, in a tin case.

A pemmican tin, painted lead colour, and marked "E." (Erebus) in black. From its size it must have contained 20 lb. or 22 lb.

Two yellow glass beads, a glass seal with symbol of Freemasonry.

A 4-inch block, strapped, with copper hook and thimble, probably for the boat's sheet.

Relics seen in lat. 69° 09′ N., long. 99° 24′ W., not brought away, 30th of May, 1859 :—

A large boat, measuring 28 ft. in extreme length, 7ft. 3 in. in breadth, 2 ft. 4 in. in depth. The markings on her stem were —"XXI. W. Con. N61., APr. 184." It appears that the fore part of the stem has been cut away, probably to reduce weight, and part of the letters and figures removed. An oak sledge under the boat, 23 ft. 4 in. long, and 2 ft. wide; 6 paddles, about 60 fathoms of deep-sea lead line, ammunition, 4 cakes of navy chocolate, shoemaker's box with implements complete, small quantities of tobacco, a small pair of very stout shooting boots, a pair of very heavy iron-shod knee boots, carpet boots, sea boots and shoes—in all seven or eight pairs; two rolls of sheet lead, elm tingles for repairing the boat, nails of various sizes for boat, and sledge irons, three small axes, a broken saw, leather cover of a sextant case, a chain-cable punch, silk handkerchiefs (black, white, and coloured), towels, sponge, tooth-brush, hair comb, a macintosh, gun cover (marked in paint "A. 12"), twine, files, knives; a small worsted-work slipper, lined with calfskin, bound with red riband; a great quantity of clothing, and a wolfskin robe; part of a boat's sail of No. 8 canvas, whale-line rope with yellow mark, and white line with red mark; 24 iron stanchions, 9½ inches high, for supporting a weather cloth round the boat; a stanchion for supporting a ridge pole at a height of 3 ft. 9 in. above the gunwale.

Relics found about Ross Cairn, on Point Victory, May and June, 1859, brought away :—

A 6-inch dip circle by Robinson, marked I 22. A case of

medicines, consisting of 25 small bottles, canister of pills, ointment, plaster, oiled silk, &c. A 2-foot rule, two joints of the cleaning rod of a gun, and two small copper spindles, probably for dog-vanes of boats. The circular brass plate broken out of a wooden gun-case, and engraved "C. H. Osmer, R.N." The field glass and German silver top of a 2-foot telescope, a coffee canister, a piece of a brass curtain rod. The record tin and the record, dated 25th of April, 1848. A 6-inch double frame sextant, on which the owner's name is engraved, "Frederick Hornby, R.N."

Found in a small cairn on the south side of Back Bay :—

A tin record case and record.

Seen about Ross Cairn, Point Victory, not brought away :—

Four sets of boat's cooking apparatus complete, iron hoops, 4 feet of a copper lightning conductor, hollow brass curtain-rod three-quarters of an inch in diameter, 3 pickaxes, 1 shovel, old canvas, a pile of warm clothing and blankets 4 feet high, 2 tin canteens stamped "89 Co., Wm. Hedges," "88 Co., Wm. Heather," and a third one not marked. A small pannikin, made on board out of a 2 lb. preserved-meat tin, and marked "W. Mark;" a small deal box for gun wadding, the heavy iron work of a large boat, part of a canvas tent, part of an oar sawed longitudinally and a blanket nailed to its flat side, three boat-hook staves, strips of copper, a 9-inch single block strapped, a piece of rope and spunyarn. Among the clothing was found a stocking marked "W," green, and a fragment of one marked "W. S."

Relics obtained at the Northern Cairn, near Cape Felix, May, 1859 :—

Fragments of a boat's ensign, metal lid of a powder-case, two eye pieces of sextant tubes, brass button; worsted glove, colours red, white, and blue; bung-stave of a marine's water keg or bottle, brass ornaments to a marine's shako; brass screw for screwing down lid, also a copper hinge of the lid of powder-case; a few patent wire cartridges containing large shot; part of a pair of steel spectacles, glass being replaced by wood, having a narrow slit in it; two small rib bones, probably out of salt pork; six or eight packets of needles; small flannel cartridge containing an ounce of damaged powder; a small, roughly made copper apparatus for cooking; some brimstone matches. Piece of white paper folded up found in the North Cairn, two pike-heads, narrow strip of white paper, found under one of the tent places; their tent places were within a few yards of the cairn.

Beside a small cairn, about three miles north of Point Victory, was a pickaxe, with broken handle; brought away an empty tea or coffee canister.

Articles noticed about the North Cairn, not brought away :—

Fragments of two broken bottles, several pieces of broken basins or cups, blue and white delfware, hoops of marine's water keg, small iron hoops, fragments of white line, spun yarn, canvas, and twine; three small canvas tents, under which lay a bearskin and fragments of blankets; two blanket frocks, several old mits, stockings, gloves, pilot cloth and box cloth jackets and trousers, large shot, piece of tobacco and broken pipe, metal part of powder-case, top of tin canister, marked "cheese," preserved-potato tin, feathers of ptarmigan, and salt-meat bones.

Seen near Cape Maria Louisa :—

Part of a drift tree, white spruce fir, 18 feet long, 10 inches in diameter; it appeared to have but recently (i. e. since thrown on the coast) been sawed longitudinally down the centre, and one-half of it removed.

Relics obtained from the Boothian Esquimaux, near the Magnetic Pole, in March and April, 1859 :—

Seven knives made by the natives out of materials obtained from the last expedition, one knife without a handle, one spear-head and staff (the latter has broken off), two files; a large spoon or scoop, the handle of pine or bone, the bowl of musk-ox horn; six silver spoons and forks, the property of Sir John Franklin, Lieutenants H. D. Vescomte and Fairholme, A. M'Donald, Assistant-Surgeon, and Lieutenant E. Couch (supposed from the initial letter T and crest a lion's head); a small portion of a gold watch-chain, a broken piece of ornamental work apparently silver gilt, a few small naval and other metal buttons, a silver medal obtained by Mr. M'Donald as a prize for superior attainments at a medical examination in Edinburgh April, 1838; some bows and arrows, in which wood, iron, or copper has been used in the construction—of no other interest.

Remarks upon these Articles.

The spear-staff measures 6 feet 3 inches in length, and appears to have been part of a light boat's gunwale: it measured (before being partially rounded to adapt it to its present use) about $1\frac{1}{2}$ by $1\frac{3}{8}$ inches, is made of English oak, and upon the side has been painted white over green. The spear-head is of steel, riveted to two pieces of hoop, with bone between, and lashed on to the staff. The rivets are of copper nails. The native who sold it said he himself got it from the boat in the Fish River. Another spear of the same kind was seen. The knives are made either of iron or steel, riveted to two strips of hoop, between which the handle of wood is inserted, and rivets passed through, securing them together.

The rivets are almost all made out of copper nails, such as would be found in a copper-fastened boat, but those which have been examined do not bear the Government mark. It is probable that most of the boats of the 'Erebus' and 'Terror' were built by contract, and therefore would not have the broad arrow stamped upon their iron and copper work. One small knife appears to have been a surgical instrument. A large knife obtained in April bears some marking, such as a sword or a cutlass might have. The man who sold it said he bought it from another, who picked it up on the land where the ship was driven ashore by the ice, and where the white people had thrown it away; it was then about as long as his arm. This was the first information he received of one of the ships having drifted on shore. One knife and one file are stamped with the broad arrow. The handles are variously composed of oak, ash, pine, mahogany, elm, and bone. The spoons and forks were readily sold for a few needles each, also the buttons, which they wore as ornaments on their dresses. Bows and arrows were readily exchanged for knives. Previously to the stranding on the neighbouring shore of the last expedition these people must have been almost destitute of wood or iron. Some of them had even got only bone knives and spear-points. Some of their sledges were seen, consisting of two rolls of sealskin, flattened and frozen, to serve as runners, and connected together by cross bars of bones. Many more knives, bows, and buttons, similar to those brought away, might have been obtained, but no personal or important relics.

Seen in a Snow Hut in lat. 70½° deg. N., 20th of April, 1859, not brought away:—

Two wooden shovels, one of them made of mahogany board, some spear-handles and a bow of English wood, a deal case which might have served for a telescope or barometer. Its external dimensions were:—length, 3 ft. 1 in.; depth, 3½ in.; width, 9 in.; two brass hinges remained attached to it.

Relics obtained from the Esquimaux near Cape Norton, upon the East Coast of King William Island, in May, 1859:—

Two tablespoons; upon one is scratched "W. W.," on the other "W. G.;" these bear the Franklin crest; two table forks, one bearing the Franklin crest, the other is also crested, probably Captain Crozier's; silversmith's name is "I. West;" two teaspoons, one engraved "A. M. D." (A. M'Donald), the other bears the Fairholme crest and motto; handle of a dessert knife, into which had been inserted a razor (since broken off) by Millikin, Strand; buttons, wood and iron, were here in abundance, but as enough of these had already been obtained no more were purchased.

Taken out of some deserted snow-huts near here, some scraps of different kinds of wood, such as could not be obtained from a boat —teak or African oak.

Found lying about the skeleton, 9 miles eastward of Cape Herschel, May, 1859 :—The tie of black silk neckerchief; fragments of a double-breasted blue cloth waistcoat, with covered silk buttons, and edged with braid; a scrap of a coloured cotton shirt, silk covered buttons of blue cloth great-coat, a small clothes-brush, a horn pocket-comb, a leathern pocket-book, which fell to pieces when thawed and dried; it contained 9 or 10 letters, a few leaves apparently blank; a sixpence, date 1831; and a half-sovereign, dated 1844.

Articles seen among the natives at Cape Norton, not purchased: —Bows made of wood, knives, uniform and plain buttons, a sledge made of two long pieces of hard wood.

From beside an Esquimaux stone-mark, on the east side of Montreal Island:—Part of a preserved-meat tin, painted red; part of the rim of some strong copper case or vessel; pieces of iron hoop, two pieces of flat iron, an iron hook bolt, a piece of sheet copper.

Articles seen about a snow-hut near Point Booth, not purchased: —Eight or 10 fir poles, varying from 5 feet to 10 feet in length, the stoutest being 2½ inches in diameter. Two wooden snow shovels about 3½ feet long, and made of pieces of plank painted white or pale yellow; it occurred to me that the pieces of plank might have been the bottom boards of a boat. There was abundance of wood fashioned into smaller articles.

Contents of Boat's Medicine Chest:—

One bottle labelled as zinzib. R. pulv., full; ditto, spirit. rect., empty; ditto, mur. hydrarg. seven-eighths full; ditto, ol. caryphyll., one-fifth full; ditto, ipec. P. co., full; ditto, ol. menth. pip., empty; ditto, liq. ammon. fort., three-quarters full; ditto, ol. olivac., full; ditto, tinct. opii. camph., three-quarters full; ditto, vin. sem. colch., full; ditto, quarter full; ditto, calomel, full (broken); ditto, hydrarg. hit. oxyd., full; ditto, pulv. gregor., full (broken); ditto, magnes. carb., full; ditto, camphor, full; two bottles tinc. tolut., each quarter full; one bottle ipec. R. pulv., full; ditto, jalap R. pulv., full; ditto, scammon. pulv., full; ditto, quinac bisulph. empty; ditto (not labelled), tinct. opii., three-quarters full; one box (apparently) purgative pills, full; ditto, ointment, shrunk; ditto, emp. adhesiv., full; one probang, one pen wrapped up in lint, one lead pencil, one pewter syringe, two small tubes (test) wrapped up in lint, one farthing, bandages, oil silk, lint, thread.

No. IV.

GEOLOGICAL ACCOUNT OF THE ARCTIC ARCHIPELAGO,

DRAWN UP PRINCIPALLY FROM THE SPECIMENS COLLECTED BY

CAPTAIN F. L. M'CLINTOCK, R.N.,
From 1849 to 1859.

BY THE REV. SAMUEL HAUGHTON, F.R.S.,
Fellow of Trinity College, Professor of Geology in the University of Dublin, and
President of the Geological Society of Dublin.

THE map which accompanies this geological description is arranged from the specimens brought home by Captain F. L. M'Clintock, R.N., from the four Arctic Expeditions in which he served from 1848 to 1859. These specimens are all deposited in the Museum of the Royal Dublin Society, and form a more extensive and better collection of Arctic rocks and fossils than is to be found in any other museum in Europe.

It will be most convenient to describe the geology of the Arctic Islands by the formations which are to be found there, which are the following:—

1. The Granitic and Granitoid Rocks.
2. The Upper Silurian Rocks.
3. The Carboniferous Rocks.
4. The Lias Rocks.
5. The Superficial Deposits.

I shall describe these successive formations briefly, and add a few remarks of a theoretical character, to indicate the important inferences which may be drawn from the facts respecting them made known to us by M'Clintock's discoveries.

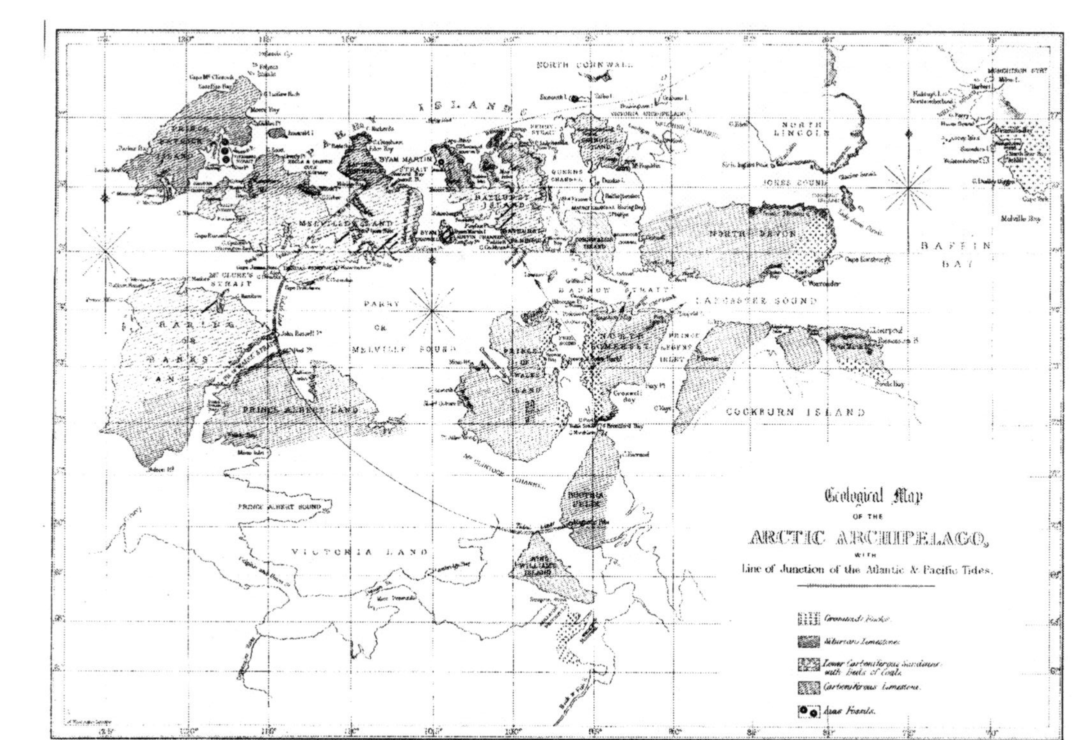

Geological Map
OF THE
ARCTIC ARCHIPELAGO,
WITH
Line of Junction of the Atlantic & Pacific Tides.
Granitic Rocks
Silurian Limestone
Lower Carboniferous Sandstone with beds of Coals
Carboniferous Limestone
Lias Fossils
NORTH CORNWALL
NORTH LINCOLN
JONES SOUND
NORTH DEVON
BAFFIN BAY
LANCASTER SOUND
BARROW STRAIT
PARRY OR MELVILLE SOUND
PRINCE PATRICK ISLAND
MELVILLE ISLAND
BATHURST ISLAND
BYAM MARTIN STRAIT
QUEENS CHANNEL
M'CLURE'S STRAIT
BARING OR BANKS LAND
PRINCE ALBERT LAND
PRINCE ALBERT SOUND
VICTORIA LAND
PRINCE OF WALES ISLAND
NORTH SOMERSET
PRINCE REGENT INLET
COCKBURN ISLAND
BOOTHIA FELIX
KING WILLIAM ISLAND
Mc CLINTOCK CHANNEL

I.—*The Granitic and Granitoid Rocks.*

These rocks form a considerable part of North Greenland, on the east side of Baffin's Bay, and constitute the rock of the country at the east side of the island of North Devon, which forms a portion of the coast-line of the west of Baffin's Bay, and the north side of the entrance into Lancaster Sound.

1. *Whale Fish Islands,* lat. 69° N., are composed of a very fine-grained, flaggy, black mica schist, composed of black mica in very small plates, occasionally putting on a hornblendic lustre, and minute grains of quartz interstratified with the mica. The softer varieties are cut by the natives into grissets and cooking utensils of various shapes, some of which resemble the cambstones found in Ireland, which are made from a kind of potstone, abundant in parts of the County Donegal.

2. *Upernavik,* lat. 72° N., Greenland.—This district is famous for the occurrence of large quantities of plumbago, which is found in a metamorphic rock of the following character. Fine-grained, amorphous, granitoid rock, composed of minute particles of grey quartz; a honey-coloured felspar of waxy lustre, of unknown composition; minute particles of red semitransparent garnet, of conchoidal fracture; and small particles, with occasional large nests, of plumbago. The plumbago occurs both amorphous, and in long acicular crystals. Sometimes the rock becomes of coarser texture and more crystalline, and the yellow colour of the felspar gives place to a greenish tinge; and it sometimes also becomes a felspar of perfect cleavage, semitransparent, and white. The dodecahedral crystals of garnet reach the diameter of one inch.

The general character of the rocks near Upernavik is different from that of the rock in which the plumbago is found; they consist of a fine-grained black mica schist, with very little felspar or quartz, and intersected by thin veins of elvan composed of quartz and white felspar. The cooking utensils of the natives are made from this fine schist, in preference to any other description of rock.

3. *Woman's Islands.*—These islands, off the west coast of Greenland, are composed of a garnetiferous mica slate, formed of black mica in layers, with alternating plates composed of white felspar and quartz, and filled with fine garnets, rose-coloured, vitreous in fracture, and transparent.

4. *Cape York*, lat. 76° N., Greenland.—This cape is composed of a fine-grained granite, consisting of quartz, white felspar, with minute specks of a black mineral, of pitchy lustre, composition not yet determined.

5. *Wolstenholme and Whale Sounds*, lat. 77° N., Greenland.—At Wolstenholme Sound the granitoid rocks of Greenland become converted into mica slate and actinolite slate of a remarkable character. The mica slate is composed of large plates of an intimate mixture of black and white mica, the chemical examination of which will doubtless prove of interest. These plates of mica are separated by bands of pure white felspar. The actinolite slate is dark green, and formed by an almost insensible gradation from the mica slate. In the low ground between Wolstenholme and Whale Sounds, the granitic rocks cease, and are covered by deposits of fine red gritty sandstone, of a banded structure, and a remarkable coarse white conglomerate. The boundary between these formations is also marked by the development of masses of dolerite and clayey basalt.

6. *Carey's Islands*, 76° 40′ N., Greenland, lie to the westward of Wolstenholme Sound, and are composed of a remarkable gneissose mica schist, formed of successive thin layers of quartz granules, containing scarcely any felspar, and layers of jet black mica, with

occasional facets of white mica. This mica schist passes into a white gneiss, composed of quartz, white felspar, and black mica, penetrated by veins, coarsely crystallised, of the same minerals. Yellow and white sandstones are also found in small quantity on the islands, reposing upon the granitoid rocks.

7. *Capes Osborn and Warrender*, lat. 74° 30′ N., North Devon.—The granitoid rocks between these two capes are composed of graphic granite, consisting of quartz (grey) and white felspar; this graphic granite passes into a laminated gneiss, consisting of layers of black mica and white translucent felspar, sparingly mixed with quartz; with the gneiss are interstratified beds of garnetiferous mica slate, consisting of quartz, pale greenish white felspar, black and white mica in minute spangles, and crystals of garnet, rose-coloured, disseminated regularly through the mass. Quartziferous bands of epidotic hornstone occur with the foregoing beds; and the whole series is overlaid by red sandstones, of banded structure, which bear a striking resemblance to those that overlie the granitoid beds of Wolstenholme Sound.

8. *North Somerset.*—The granitoid rocks are found again on the west side of the island of North Somerset, where they form the eastern boundary of Peel Sound. Boulders of granite are found at a considerable distance (100 miles) to the north-eastward of the rock *in situ*, as at Port Leopold, Cape Rennell, &c. The general character of the granitic rocks in the north and west of North Somerset are thus described by Captain M'Clintock:—

"Near Cape Rennell we passed a very remarkable rounded boulder of gneiss or granite; it was 6 yards in circumference, and stood near the beach, and some 15 or 20 yards above it; one or two masses of rounded gneiss, although very much smaller, had arrested our attention at Port Leopold, as then we knew of no such formation nearer than Cape Warrender, 130 miles to the north-east; subsequently we found it to commence *in situ* at Cape Granite, nearly 100 miles to the south-west of Port Leopold.

"The granite of Cape Warrender differs considerably from that of North Somerset; the former being a graphic granite, composed of grey quartz and white felspar, the quartz predominating; while the latter, or North Somerset granite, is composed of grey quartz, red felspar, and green chloritic mica, the latter in large flakes; both the granite and gneiss of North Somerset are remarkable for their soapy feel." *

To the east of Cape Bunny, where the Silurian limestone ceases, and south of which the granite commences, is a remarkable valley called Transition Valley, from the junction of sandstone and limestone that takes place there. The sandstone is red, and of the same general character as that which rests upon the granitoid rocks at Cape Warrender and at Wolstenholme Sound. Owing to the mode of travelling, by sledge on the ice, round the coast, no information was obtained of the geology of the interior of the country, but it appears highly probable that the granite of North Somerset, as well as that of the other localities mentioned, is overlaid by a group of sandstones and conglomerates, on which the Upper Silurian limestones repose directly. A low sandy beach marks the termination of the valley northwards, and on this beach were found numerous pebbles, washed from the hills of the interior, composed of quartzose sandstone, carnelian, and Silurian limestone. The accompanying sketch was made by Captain M'Clintock, on the spot, in 1849, and afterwards finished by Lieutenant Browne. It represents the island called Cape Bunny, which forms the eastern headland of the entrance of the now famous Peel Sound, down which the 'Erebus' and 'Terror' sailed, three years before it was

---

* Journal of the Royal Dublin Society, 1857.

No. IV.

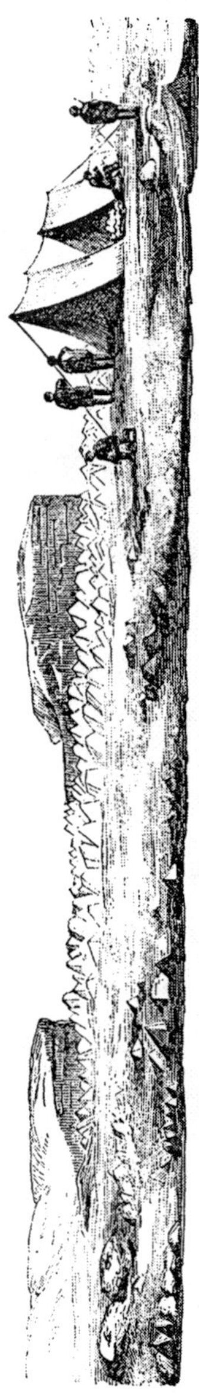
Cape Bunny, Peel Sound.

visited by Sir James C. Ross and Lieutenant M'Clintock, in their first sledge journey on the ice. Cape Granite is the northern boundary of the granite, which retains the same character as far as Howe Harbour. It is composed of quartz, red felspar, and dark green chlorite; and is accompanied with gneiss of the same composition. I have in my possession a specimen of this granite, found as a pebble at Graham Moore Bay, Bathurst Island, S.W., a locality 135 knots distant from Cape Granite, to the N.W.

9. *Bellot's Straits*, lat. 72° N., separate North Somerset from Boothia Felix. The 'Fox' Expedition wintered here in 1858, and had abundant means of ascertaining the geological structure of the neighbourhood. The junction of the granitoid and Silurian rocks occurs in these straits, the low ground to the east being horizontal beds of Silurian limestone, while on the west the granite hills of West Somerset rise to a height of 1600 feet above the narrow straits. The granite here is of three varieties.

α. Blackish grey, fine-grained, gneissose granite, composed of quartz, white felspar, and large quantities of fine grains and flakes of hornblende, passing into black mica. The gneissose beds of this granite dip 13° S.E.

β. A red granite, graphic texture, composed of quartz and red felspar, coarse grained.

γ. Syenite, composed of honey-yellow felspar and hornblende, in very large crystals, the felspar passing into red and pink, and the whole rock mass penetrated by veins of the same material, but fine grained. This variety of igneous rock was met with principally at Pemmican Rock, western inlet of Bellot's Straits. Large quantities of hornblende are also met with at Leveque Harbour, Bellot's Straits, composed of facetted crystals agglutinated together into large masses, forming a crystalline hornblendic gneiss.

10. *Pond's Bay, Baffin's Bay*, lat. 72° 40′ N.—In this locality a quartziferous black mica schist underlies the Silurian limestone, and is interstratified with gneiss and garnetiferous quartz rock, all in beds, inclined 38° W.S.W. (true).

11. *Montreal Island*, mouth of the Fish River, lat. 67° 45′ N.—The granitoid rocks, which everywhere, in the Arctic Archipelago, underlie the Silurian limestone, appear at Montreal Island as a gneiss, composed of bands of felspar (pink) and quartz (¼ inch thick), separated by thin plates composed altogether of black mica; the whole rock exhibiting the phenomena of foliation in a marked degree.

The east side of King William's Island, though composed of Silurian limestone like the rest of the island, is strewed with boulders of black and red micaceous gneiss, like that of Montreal Island, and black metamorphic clay slate, in which the crystals of mica (qu. Ottrelite) are just commencing to be developed. It is probable that the granitoid rocks appear at the surface somewhat to the eastward of this locality.

12. *Prince of Wales' Island*, west of Peel Sound.—The granitoid rocks extend across Peel Sound into Prince of Wales' Island, in the form of a dark syenite, composed of quartz, greenish white felspar passing into

yellow, and hornblende. This rock is massive and eruptive at Cape M'Clure, lat. 72° 52′ N., and occasionally gneissose, as at lat. 72° 13′ N. Between these two points, at lat. 72° 37′ N., a limestone bluff occurs containing the characteristic Silurian fossils, and is succeeded at 72° 40′ by a ferruginous limestone, bright red, and a few beds of fine red sandstone, like those observed by M'Clintock at Transition Valley, North Somerset. The entire western portion of Prince of Wales' Land is composed of Silurian limestone, which in the extreme west, at Cape Acworth, becomes chalky in character and non-fossiliferous, resembling the peculiar Silurian limestone found on the west side of Boothia Felix.

II.—*The Silurian Rocks.*

The Silurian rocks of the Arctic Archipelago rest everywhere directly on the granitoid rocks, with a remarkable red sandstone, passing into coarse grit, for their base. This sandstone is succeeded by ferruginous limestone, containing rounded particles of quartz, which rapidly passes into a fine greyish green earthy limestone, abounding in fossils, and occasionally into a chalky limestone, of a cream colour, for the most part devoid of fossils. The average dip of the Silurian limestone varies from 0° to 5° N.N.W., and it forms occasionally high cliffs, and occasionally low flat plains, terraced by the action of the ice as the ground rose from beneath the sea. The general appearance of the rocks is similar to the Dudley limestone, and would strike even an observer who was not a geologist. This resemblance to the Upper Silurian beds extends to the structure of the rocks on the large scale. Alternations of hard limestone and soft shale, so characteristic of the Upper Silurian beds of England and America, arranged in horizontal layers, give to the cliffs around Port

Leopold the peculiar appearance which has been described by different Polar navigators as "buttress-like," "castellated;" this appearance is produced by the unequal weathering of the cliff, which causes the hard limestone to stand out in bands. Excellent sketches of this remarkable appearance, drawn by Lieutenant Beechey, are figured at page 35 of Parry's First Voyage, 'Hecla' and 'Griper,' 1819-20. The western side of King William's Island (now, alas! invested with so sad an interest) is a good example of the low terraced form which the limestone rocks assume at times.

The following lists contain the names of the principal fossils brought home by Captain M'Clintock:—

No. I. GARNIER BAY (Lat. 74° N.; Long. 92° W.).

1. *Cyathophyllum helianthoides*, several specimens.
2. *Heliolites porosa*. Garnier Bay. Another specimen from near Cape Bunny.
3. Specimens of carnelian, gneiss, chalcedony, &c. &c., from the shingle near Cape Bunny.
4. *Cromus Arcticus*, several specimens.
5. *Atrypa phoca* (Salter).
6. *Atrypa reticularis*.
7. Brachiopoda on slab (various).
8. Cyathophyllum.
9. *Columnaria Sutherlandi* (Salter). Several specimens.

No. II. PORT LEOPOLD (Lat. 73° 50′ N.; Long. 90° 15′ W.).

1. Limestone containing numerous fossils of the Upper Silurian type: *Calamopora Gothlandica*, Goldf. *Rhynchonella cuneata?* Dalm. *Cyathophyllum*, sp.
2. Dark earthy limestone, containing multitudes of the *Loxonema M'Clintocki*, as casts—1100 feet above sea-level on North-east Cape.
3. Fine specimens of selenite from shaly beds in cliff.
4. Fibrous gypsum from same.

No. III. GRIFFITH'S ISLAND (Lat. 74° 35′ N.; Long. 95° 30′ W.).

1. Beautiful specimens of the *Cromus Arcticus*. Pl. VI. Fig. 5, Journ. R. D. S., Vol. I.
2. *Orthoceras Griffithi*. Pl. V. Fig. 1, Journ. R. D. S., Vol. I.
3. An Orthoceras with lateral siphuncle, and simple circular outline of septa.

4. *Loxonema Rossi.* Pl. V. Figs. 6, 8, 9, 10, 11, Journ. R. D. S., Vol. I.
5. Numerous specimens of crinoidal limestone.
6. *Strophomena Donnetti* (Salter). Sutherland's Voyage; Pl. V. Figs. 11, 12.
7. *Atrypa phoca* (Salter). Pl. V. Figs. 3, 4, 7, Journ. R. D. S., Vol. I.; and a ribbed Atrypa, not identified with European species, and undescribed.
8. An undescribed bryozoan Zoophyte. Pl. VII. Fig. 6, Journ. R. D. S., Vol. I.
9. *Calophyllum Phragmoceras* (Salter). Sutherland; Pl. VI. Fig. 4.
10. *Syringopora geniculata.*
11. An undescribed species of *Macrocheilus.*

No. IV. BEECHEY ISLAND (Lat. 74° 40′ N.; Long. 92° W.).

1. Orthoceras (species).
2. Great multitudes of *Atrypa phoca,* forming, in fact, a dark-coloured earthy Atrypa limestone.
3. With these were associated many species of Loxonema, sometimes so abundant as to form a pale pink and whitish Loxonema limestone.
4. A species of ribbed Atrypa.
5. Crinoidal limestone in abundance.
6. *Syringopora reticulata.*
7. *Calophyllum phragmoceras* (Salter). Sutherland; Pl. VI. Fig. 4.
8. *Cyathophyllum cæspitosum.*
9. *Cyathophyllum articulatum* (Edwardes and Haime).
10. *Calamopora Gothlandica.*
11. *Calamopora alveolaris.*
12. *Favistella Franklini* (Salter). Sutherland; Pl. VI. Fig. 3.
13. *Clisiophyllum Salteri.* Sutherland; Pl. VI. Fig. 7.
14. *Cyathophyllum* (species).
15. *Loxonema Salteri,* described by Mr. Salter in Sutherland's 'Voyage to Wellington Channel;' Pl. V. Fig. 19.

This is a fine slab of limestone, almost altogether composed of the remains of *Loxonema Salteri* and *Atrypa phoca.* It appears to have been quietly deposited at the bottom of a deep submarine depression, swarming with Pyramidellidæ and deep-water Brachiopoda. The physical conditions indicated by the fossils are also rendered probable by the rock itself, which consists of fine grey limestone, subcrystalline, and intimately blended with the finest and most delicate description of mud, such as could only be found where the water was deep, and all currents far removed.

No. V. CORNWALLIS ISLAND, Assistance Bay (Lat. 74° 40′ N.; Long. 94° W.).

1. *Orthoceras Ommaneyi* (Salter). Sutherland; Pl. V. Figs. 16, 17.
2. *Pentamerus conchidium* (Dalman). Sutherland; Pl. V. Figs. 9, 10.

3. Pentamerus limestone.
4. *Cromus Arcticus.* Journ. R. D. S., Vol. I. Pl. VI.
5. *Cardiola Salteri.* Pl. VII. Fig. 5. Journ. R. D. S., Vol. I.
6. *Syringopora geniculata.*

No. VI. CAPE YORK, Lancaster Sound (Lat. 73° 50′ N.; Long. 87° W.).

A specimen of the same fossil coral which I have named, doubtfully, from Beechey Island, as Favosites or *Calamopora Gothlandica;* it is not impossible, however, that it is not a Calamopora at all, but a species of Chætetes.

No. VII. POSSESSION BAY, South Entrance into Lancaster Sound (Lat. 73° 30′ N.; Long. 77° 20′ W.).

Specimens of brown earthy limestone, with a fetid smell when struck with a hammer; resembles closely the limestone of Cape York, Lancaster Sound.

No. VIII. DEPOT BAY, Bellot's Straits (Lat. 72° N.; Long. 94° W.).

1. *Maclurea* sp.
2. *Cyathophyllum helianthoides* (Goldfuss).

The limestone at this locality is white and saccharoid, with large rhombohedral crystals of calcspar.

*No. IX. CAPE FARRAND, East side of Boothia (Lat. 71° 38′; Long. 93° 35′ W.).

1. *Atrypa phoca* (Salter). Sutherland; Pl. V. Fig. 3.
2. *Loxonema Rossi.* Journ. R. D. S., Vol. I. Pl. V.
3. *Atrypa* (ribbed sp.).
4. *Calamopora Gothlandica* (Goldfuss).
5. *Cyrtoceras* sp.

The rock at this locality is a grey mud limestone.

No. X. WEST SHORE OF BOOTHIA (Lat. 70° to 71° N.), containing the Magnetic Pole.

1. *Atrypa phoca* (Salter).
2. *Loxonema Rossi.* Journ. R. D. S., Vol. I. Pl. V.
3. *Favistella Franklini* (Salter). Journ. R. D. S., Vol. I. Pl. XI.
4. *Loxonema Salteri.* Sutherland; Pl. V. Fig. 18.

The cream-coloured chalky limestone found on the west side of Prince of Wales' Island here occurs, and is generally destitute of fossils, like that of Prince of Wales' Land.

*No. XI. FURY POINT (Lat. 72° 50′ N.; Long. 92° W.).

1. *Cromus Arcticus.* Journ. R. D. S., Vol. I. Pl. VI.
2. *Maclurea* sp.

---

* Collected by Dr. Walker, surgeon to the 'Fox' Expedition.

3. *Mya rotundata* (?).
4. *Stromatopora concentrica.*
5. *Cyathophyllum helianthoides* (Goldfuss).
6. *Petraia bina.*
7. *Calamopora Gothlandica* (Goldfuss).
8. *Favosites megastoma* (?).
9. *Cyathophyllum cæspitosum.*
10. *Favistella Franklini* (Salter). Sutherland; Pl. VI. Fig. 3.
11. *Strephodes Austini* (Salter). Sutherland; Pl. VI. Fig. 6.
12. *Atrypa phoca* (Salter.)

The limestone here is of the same grey earthy aspect as at Beechey Island and Port Leopold.

*No. XII. PRINCE OF WALES' LAND (Lat. 72° 38′ N.; Long. 97° 15′ W.).

1. *Cyathophyllum* sp.
2. *Calamopora Gothlandica* (Goldfuss).
3. *Stromatopora concentrica.*

These fossils occur in grey earthy limestone, near its junction with the red arenaceous limestone already described.

No. XIII. WEST COAST OF KING WILLIAM'S ISLAND.

1. *Loxonema Rossi.* Journ. R. D. S., Vol. I. Pl. V.
2. *Catenipora escharoides.*
3. *Orthoceras* sp.
4. *Maclurea* sp.
5. *Atrypa* sp.
6. *Syringopora geniculata.*
7. *Clisiophyllum* sp.
8. *Orthis elegantula.*

III.—*The Carboniferous Rocks.*

The Upper Silurian limestones already described are succeeded by a most remarkable series of close-grained white sandstones, containing numerous beds of highly bituminous coal, and but few marine fossils. In fact, the only fossil shell found in these beds, so far as I know, in any part of the Arctic Archipelago, is a species of ribbed *Atrypa*, which I believe to be identical with the *Atrypa fallax* of the carboniferous slate of Ireland. These sandstone beds are succeeded by a series of blue limestone beds, containing an abundance of the marine

* Collected by Captain Allen Young.

shells commonly found in all parts of the world where the carboniferous deposits are at all developed. The line of junction of these deposits with the Silurians on which they rest is N.E. to E.N.E. (true). Like the former they occur in low flat beds, sometimes rising into cliffs, but never reaching the elevation attained by the Silurian rocks in Lancaster Sound.

The following lists contain the principal fossils and specimens presented to the Royal Dublin Society by Captain M'Clintock and by Captain Sir Robert M'Clure.

Coal, sandstone, clay ironstone, and brown hematite, were found along a line stretching E.N.E. from Baring Island, through the south of Melville Island, Byam Martin's Island, and the whole of Bathurst Island. Carboniferous limestone, with characteristic fossils, was found along the north coast of Bathurst Island, and at Hillock Point, Melville Island.

I have marked on the map the coal-beds of the Parry Islands, which appear to be prolonged into Baring Island, as observed by Captain M'Clure. The discovery of coal in these islands is due to Parry, but the evidence of the extent and quantity in which it may be found was obtained during the expeditions of Austin and Belcher. In addition to the localities surveyed by himself, Captain M'Clintock has given me specimens of the coal found at other places by other explorers; and it is from a comparison of all these specimens that I have ventured to lay down the outcrop of the coal-beds, which agrees remarkably well with the boundary of the formations laid down from totally different data.

No. I. HILLOCK POINT, Melville Island (Lat. 76° N.; Long. 111° 45′ W.).

Productus sulcatus. Journ. R. D. S., Vol. I. Pl. VII. Figs. 1, 2, 3, 4, 7.
Spirifer Arcticus. Journ. R. D. S., Vol. I. Pl. IX.

No. II. BATHURST ISLAND, North Coast, Cape Lady Franklin (?) (Lat. 76° 40′ N.; Long. 98° 45′ W.).

Spirifer Arcticus. Journ. R. D. S., Vol. I. Pl. IX. Fig. 1.
Lithostrotion basaltiforme.

*No. III. BALLAST BEACH, Baring Island (Lat. 74° 30′ N.; Long 121° W.).

1. Wood fossilized by brown hematite; structure quite distinct.
2. Cone of the spruce fir, fossilized by brown hematite.

No. IV. PRINCESS ROYAL ISLANDS, Prince of Wales' Strait, Baring Island (Lat. 72° 45′ N.; Long. 117° 30′ W.).

1. Nodules of clay ironstone, converted partially into brown hematite.
2. Native copper in large masses, procured from the Esquimaux in Prince of Wales' Strait.
3. Brown hematite, pisolitic.
4. Greyish-yellow sandstone, same as Cape Hamilton and Byam Martin's Island.
5. *Terebratula aspera* (Schlotheim). Journ. R. D. S., Vol. I. Pl. XI. Fig. 4.

This interesting brachiopod was found in limestone by Captain M'Clure, at the Princess Royal Islands, in the Prince of Wales' Strait, between Baring Island and Prince Albert Land. I have no hesitation in pronouncing it to be identical with Schlotheim's fossil, which is found in the greatest abundance at Gerolstein, in the Eifel. Banks' Land, or Baring Island, is composed of sandstone, similar to that at Byam Martin's Island, and at the Bay of Mercy. This sandstone contains beds of coal, apparently the continuation of the well-known coal-beds of Melville Island. It is a remarkable fact, that these carboniferous sandstones *underlie* beds of undoubtedly the carboniferous limestone type, and that at Byam Martin's Island, where fossils are found in this sandstone, they are allied to *Atrypa fallax* and other forms characteristic of the lower sandstones of the carboniferous epoch. It is, therefore, highly probable that the coal-beds of Melville Island are very low down in the series, and do not correspond in geological position with the coal-beds of Europe, which rest on the summit of the carboniferous beds. It is interesting to find at Princess Royal Island,

* These specimens are "*Drift*," but are mentioned here, as they were found on the carboniferous sandstone area.

where, from the general strike of the beds, we should expect to find the Silurian limestone underlying the coal-bearing sandstones, that this limestone does occur, and contains a fossil, *T. aspera,* eminently characteristic of the Eifelian beds of Germany, which form, in that country, the Upper Silurian strata.

No. V. CAPE HAMILTON, Baring Island (Lat. 74° 15′ N.; Long. 117° 30′ W.).

1. Greyish-yellow sandstone, like that found *in situ* in Byam Martin's Island.
2. *Coal.*—The coal found in the Arctic regions, excepting that brought from Disco Island, West Greenland, which is of tertiary origin, presents everywhere the same characters, which are somewhat remarkable. It is of a brownish colour and lignaceous texture, in fine layers of brown coal and jet-black glossy coal interstratified in delicate bands not thicker than paper. It has a woody ring under the hammer, recalling the peculiar clink of some of the valuable gas coals of Scotland. It burns with a dense smoke and brilliant flame, and would make an excellent gas coal; and, in fact, it resembles in many respects some varieties of the coal which has acquired such celebrity in the Scotch and Prussian law-courts, under the title of the Torbane Hill mineral.

No. VI. CAPE DUNDAS, Melville Island (Lat. 74° 30′ N.; Long 113° 45′ W.).

Fine specimens of coal.

No. VII. CAPE SIR JAMES ROSS, Melville Island (Lat. 74° 45′ N.; Long. 114° 30′ W.).

Sandstone passing into blue quartzite.

No. VIII. CAPE PROVIDENCE, Melville Island (Lat. 74° 20′ N.; Long. 112° 30′ W.).

A specimen of crinoidal limestone, apparently similar to that occurring in Griffith's Island, from which, however, it could not have been brought by the present drift of the floating ice, as the set of the currents is constant from the west. If brought to its present position by ice, it must have been under circumstances differing considerably from those now prevailing in Barrow's Strait.

Yellowish-grey sandstone.

Clay ironstone passing into pisolitic hematite.

No. IX. WINTER HARBOUR, Melville Island (Lat. 74° 35′ N.; Long. 110° 45′ W.).

Fine yellow and grey sandstone.

No. X. BRIDPORT INLET, Melville Island (Lat. 75° N.; Long. 109° W.).

Coal, with impressions of Sphenopteris.
Ferruginous spotted white sandstone.
Clay ironstone, passing into brown hematite.

No. XI. SKENE BAY, Melville Island (Lat. 75° N.; Long. 108° W.).

Bituminous coal, with finely divided laminæ, associated with brown crystalline limestone, with cherty beds, and grey-yellowish sandstone, passing into brownish-red sandstone.

No. XII. HOOPER ISLAND, Liddon's Gulf, Melville Island (Lat. 75° 5′ N.; Long. 112° W.).

Nodules of clay ironstone, very pure and heavy, associated with ferruginous fine sandstone and coal of the usual description.

The hill-tops and sides along the south shore of Liddon's Gulf, and as far as Cape Dundas, are generally bare, composed of frozen mud, arising from the disintegration of shale, the annual dissolving snows washing them down and giving them a rounded form. The southern slopes generally support vegetation. Fragments of coal are very frequently met with, and at the mouth of a ravine on the south shore of Liddon's Gulf there is abundance, of very good quality; it contains a considerable quantity of pyrites or bisulphuret of iron.

No. XIII. BYAM MARTIN'S ISLAND (Lat. 75° 10′ N.; Long. 104° 15′ W.).

Yellowish-grey sandstone, *in situ*, containing a ribbed *Atrypa*, allied to the *A. primipilaris* of V. Buch, and the *A. fallax* of the carboniferous rocks of Ireland.
Reddish limestone, with broken fragments of shells, of the same description of brachiopod as the last.
Coal of the usual description.
Fine-grained red sandstone, passing into red slate.
Scoriaceous hornblendic trap (boulders).

The sandstone of Byam Martin's Island is of two kinds—one red, finely stratified, passing into purple slate, and very like the red sandstone of Cape Bunny, North Somerset, and some varieties of the red sandstone and slate found between Wolstenholme Sound and

Whale Sound, West Greenland, lat. 77° N. The other sandstone of Byam Martin's Island is fine, pale-greenish, or rather greyish-yellow, and not distinguishable in hand specimens from the sandstone of Cape Hamilton, Baring Island. It contains numerous shells and casts of a terebratuliform brachiopod, closely allied to the *Terebratula primipilaris* of Von Buch, found abundantly at Gerolstein in the Eifel. On the whole, I incline to the opinion that the sandstones, limestone, and coal of Byam Martin's Island, and the corresponding rocks of Melville Island, Baring Island, and Bathurst Island, are low down in the Carboniferous System, and that there is in these northern coal-fields no subdivision into red sandstone, limestone, and coal-measures, such as prevail in the west of Europe. If the different points where coal was found be laid down on a map, we have in order, proceeding from the south-west—Cape Hamilton, Baring Island; Cape Dundas, Melville Island, south; Bridport Inlet and Skene Bay, Melville Island; Schomberg Point, Graham Moore Bay, Bathurst Island; a line joining all these points is the outcrop of the coal-beds of the south of Melville Island, and runs E.N.E. At all the localities above mentioned, and, indeed, in every place where coal was found, it was accompanied by the greyish-yellow and yellow sandstone already described, and by nodules of clay ironstone, passing into brown hematite, sometimes nodular and sometimes pisolitic in structure.

No. XIV. GRAHAM MOORE'S BAY, Bathurst Island (Lat. 75° 30′ N.; Long. 102° W.).

Coal of the usual quality.

At Cape Lady Franklin, and at many other localities along the north shore of Bathurst Island, carboniferous fossils in limestone, clay ironstone balls passing into brown hematite, cherty limestone, and earthy fossiliferous limestone, with the same species of *Atrypa* as at

Byam Martin's Island, were found in abundance by Sherard Osborn, Esq., Commander of H.M.S. 'Pioneer,' in whose journal the following note respecting them may be found:—

"The above collection was delivered over to Captain Sir Edward Belcher, C.B., by Commander Richards, at 2 P.M., on 7th Nov. 1853."*

It is to be hoped that they may soon be made available for the elucidation of the geology of this most interesting portion of the Arctic discoveries.

No. XV. BATHURST ISLAND, Bedford Bay (Lat. 75° N.; Long. 95° 50′ W.).

In this locality abundance of vesicular scoriaceous trap rocks were found by Captain M'Clintock; they appear to me to be the representatives of the volcanic rocks found everywhere at the commencement of the carboniferous period.

No. XVI. CORNWALLIS ISLAND, M'Dougall Bay.

1. *Syringopora geniculata.* Journ. R. D. S., Vol. I. Pl. XI. Fig. 2.
2. *Cardiola Salteri.* Journ. R. D. S., Vol. I. Pl. VII. Fig. 5.

The Syringopore found at Cornwallis Island appears to be identical with the variety of the Irish carboniferous *S. geniculata,* in which the corallites are at a distance from each other somewhat exceeding their diameters, and in which the connecting tubes are about two diameters apart.

A question of very considerable geological interest is raised by the occurrence together of corals, in the same locality, of silurian and carboniferous forms.

I entertain no doubt of their being *in situ,* and occurring in the same beds, for the following reasons:—

1st. The Syringopores of Griffith's Island were found at an elevation of 400 feet above the sea, and, therefore, could not be brought by drifting ice.

2nd. The specimens were apparently of the same texture and composition as the native rock, whenever the latter was visible from under the snow.

---

* *Vide* Arctic Expeditions, 1854-55, p. 254.

3rd. I do not believe in the lapse of a long interval of time between the silurian and carboniferous deposits,—in fact, in a Devonian period.

4th. The same blending of corals has been found in Ireland, the Bas Boulonnais, and in Devonshire, where silurian and carboniferous forms are of common occurrence in the same localities.

5th. In the carboniferous beds proper of Melville Island and Bathurst Island, there were not found, so far as I am aware, any corals of the same character as those at Griffith's Island, Cornwallis Island, and Beechey Island, which could give a supply to be drifted to the latter localities in a Pleistocene sea. It is plain, from the height at which the corals were found, that, if they were brought to their present localities by ice, it must have been during the period known as Post-tertiary, as the present conditions of drift-ice in Barrow's Straits do not permit us to suppose them to have been placed where we now find them by existing causes.

The occurrence of coal-beds in such high latitudes has been speculated on by many geologists—in my opinion, not very satisfactorily; as it is very difficult to conceive how, even if the question of temperature were settled, plants even of the fern and lycopodium type could exist during the darkness of the long winter's night at Melville Island. This difficulty is increased by the facts made known to us by the discovery of ammonites and lias fossils in Prince Patrick's Island by Captain M'Clintock.

IV.—*The Lias Rocks.*

Many years ago it was asserted by Lieutenant Anjou, of the Russian navy, that ammonites had been found by him in the cliffs on the south shore of the island of New Siberia, off the north coast of Asia, in lat. 74° N. This statement, which was published in Admiral Von Wrangel's journal, attracted but little attention, until it was confirmed, as far as probability of such fossils occurring at so high a latitude is concerned, by the remarkable discovery of similar fossils by Captain M'Clintock, in lat. 76° 20′ N., at Point Wilkie, in Prince Patrick's Island.

In a paper, published by the Royal Dublin Society, in the first volume of their journal, p. 223, Captain M'Clintock thus describes the finding of these fossils:—

"After returning to Cape de Bray, we took up the provisions that the officer after whom it is called had left for us, and crossed the strait to Point Wilkie; reached it on the 14th May. This traverse was the more difficult from the great load upon our sledge, and the unfavourable state of the ice and snow. The freshly fallen snow was soft and deep, and beneath it the older snow lay in furrows across our route, hardened and polished by the winter gales and drifts, so that it resembled marble.

"On landing I found the beach low, composed of mud, with the foot-prints of animals frozen in it. A few hundred yards from the beach there are steep hills, about 150 feet in height, and upon the sides of these, in reddish-coloured limestone, casts of fossil shells abound. Inland of these, the ordinary pale carboniferous sandstone and cherty limestone reappeared. The fossils are all small, and of only a few varieties, some being ammonites, but the greater part bivalves. They differed from any I had met with before, and the rock was almost brick-red; I picked up what appeared to be fossil bone (*Ichthyosaurus?*), only part of it appearing out of the fragment of the rock.

"Point Wilkie appears to be an isolated patch of liassic age, resting upon carboniferous sandstones and limestones, with bands of chert, of the same age as the limestones and sandstones of Melville Island. The eastern shore of Intrepid Inlet is composed of this formation; while the western, rising into hills and terraces, is of the underlying carboniferous epoch. At the western side of Intrepid Inlet I found upon the ice a considerable quantity of white asbestos, but did not ascertain from whence it had been brought."

The fossils thus found *in situ*, I have no doubt, belong to the liassic period; and as their geological interest is indubitable, I offer no apology for inserting here the following description, written by me on Captain M'Clintock's return to Dublin from his third Arctic expedition.

No. I. WILKIE POINT, Prince Patrick's Land (Lat. 76° 20′ N.; Long. 117° 20′ W.).

LIAS FOSSILS.

(*a*) *Ammonites M'Clintocki.* Journ. R. D. S., Vol. I. Pl. IX. Figs. 2, 3, 4.
Monotis septentrionalis. Journ. R. D. S., Vol. I. Pl. IX. Figs. 6, 7.
Pleurotomaria, sp. Journ. R. D. S., Vol. I. Pl. IX. Fig. 8.
Cast of some Univalve. Journ. R. D. S., Vol. I. Pl. IX. Fig. 7.
Nucula, sp.

(*a*) Ammonites M'Clintocki (Haughton).—*Testâ compressâ, carinatâ, anfractibus latis, lateribus complanatis, transversim undato-costatis; costis simplicibus, juxtâ marginem interiorem levigatis; dorso carinato acuto; aperturâ sagittatâ, compressâ, antice carinatâ; septis lateribus 4-lobatis.*

This fine ammonite resembles several species common in the upper lias of the Plateau de Larzac, Sevennes, in France. It approaches *A. concavus* of the lower Oolite, but is distinguished by having only four lobes on the lateral margins of the septa, and by its showing no tendency to a tricarinated keel. The following measurements give an exact idea of its form, as compared with that of the species mentioned:—

	Diameter. Inches.	Width of last Spire. Diam.=100.	Thickness of last Spire.	Overlapping of last Spire.	Width of Umbilic.
A. M'Clintocki.	1·83	$\frac{51}{100}$	$\frac{24}{100}$	$\frac{20}{100}$	$\frac{20}{100}$
A. concavus	2·95	$\frac{50}{100}$	$\frac{24}{100}$	$\frac{19}{100}$	$\frac{16}{100}$

The principal difference here observable is in the somewhat greater size of *A. concavus*, and the larger umbilic of *A. M'Clintocki*. It certainly resembles this well-known ammonite very closely; and it appears to me difficult to imagine the possibility of such a fossil living in a frozen, or even a temperate sea.

The discovery of such fossils *in situ*, in 76° north latitude, is calculated to throw considerable doubt upon the theories of climate which would account for all past changes of temperature by changes in the relative position of land and water on the earth's surface. No attempt, that I am aware of, has ever been made to calculate the number of degrees of change possible in consequence of changes of position of land and water; and from some incomplete calculations I have myself made on the subject, I think it highly improbable that such causes could have ever produced a temperature in the sea at 76° north latitude which would allow of the existence of ammonites, especially ammonites so like those that lived at the same time in the tropical warm seas of the south of England and France, at the close of the Liassic, and commencement of the lower Oolitic period.

During the course of the same Arctic expedition in which these organic remains were found, Captain Sir Edward Belcher discovered in some loose rubble, of which a cairn was built on Exmouth Island (lat. 77° 12′ N., long. 96° W.), vertebral bones of, apparently, some liassic enaliosaurian. All doubt as to the reality of this discovery, and all idea of accounting for the occurrence of such remains by drift, must be abandoned, as the fossils found by M'Clintock were unquestionably *in situ*, and it is impossible to evade the consequences that follow to geological theory from their discovery.

Captain Sherard Osborn, also, found broken vertebræ of an ichthyosaurus, 150 feet up Rendezvous Hill, the north-west extreme of Bathurst Island: of these specimens, one lay among a mass of stone that had slipped from the N.W. face of the hill; the other was by the side of a ravine or deep watercourse on the southern face of the same elevation. I have no doubt but that they were *in situ.*

I am well aware that the question of light in the Arctic seas will be disposed of by some geologists, who will remind us that the saurians, and probably the ammonites, were endowed with a complicated optical apparatus, rendering them capable of using their eyes, not only for the distinct vision of objects differing greatly in distance, but also of using them, under widely differing conditions of light and darkness; and I readily admit the force of such observations.

But what are we to say as to the question of temperature? It was certainly necessary for an ammonite to have a sea free from ice, on which to float and bask in the pale rays of the Arctic sun; and therefore I claim a temperature for those seas, at least similar to that which now prevails in the British Islands: and I may add that the ammonite, from its habits, was essentially dependent on the temperature of the air, as well as on that of the water.

There is at present a difference of 49°·5 F. between the mean annual temperature of Point Wilkie and Dublin; and if this change of temperature be supposed to be caused by a change of the relative positions of land and water, the temperature of Dublin, or of some place on the same parallel of latitude, must be supposed to be raised to 99°·5 F.; while the temperature of the thermal equator will exceed 124°—a temperature only a few degrees below that requisite to boil an egg! I reject, without scruple, a theory that requires such a result, which must be considered as a minimum; as it is probable that the ammonite required a finer climate than that of Britain for the full enjoyment of his existence.

The theory of central heat, also, appears to me to be open to the same objection, as a mode of explaining this remarkable geological fact; for it will simply add a constant to our present climates, leaving the differences to remain, as at present, to be accounted for by latitude and distribution of land and water. The astronomical theory of Herschel, also, which would account for former changes of climate by changes in the radiating power of the sun, would only increase the temperature at each latitude, leaving the differences as at present.

The only speculation with which I am acquainted, which is capable of solving this *opprobrium geologicorum*, is the hypothesis of a change in the axis of rotation of the earth, the admission of which, as a geological possibility, is mathematically demonstrable, and which has recently had some singular evidence in its favour advanced by geologists. In 1851 I brought forward, at the Geological Society of Dublin, a case of angular fragments of granite occurring in the carboniferous limestone of the County Dublin; and explained the phenomena by the supposition of the transporting power of ice. In 1855 Professor Ramsay laid before the Geological Society of London a full and detailed theory of glaciers and ice as agents concerned in the formation of a remarkable breccia, of Permian age, occurring in the central counties of England; and still more recently the same agent has been employed by the geological surveyors of India to account for the transport of materials at geological periods long antecedent to those in which ice transport is commonly supposed to have commenced. The motion of the earth's axis would reconcile all the facts known, and it must be regarded as a geological desideratum to determine its amount and direction, and to assign the cause of such a movement. The solution of this problem I regard as quite possible.

It is well worthy of remark, that the arguments from the occurrence of coal-plants and ammonites strengthen each other; the coal-plants rendering the question of *light*, and the ammonites that of *heat*, insuperable objections to the admission of any received geological hypothesis to account for the finding of such remains, *in situ*, in latitudes so high as those of Melville Island, Prince Patrick's Island, and Exmouth Island.

V.—*The Superficial Deposits.*

The surface of the ground, where exposed, throughout the Arctic Archipelago, does not appear to be covered with thick deposits of clay or gravel, such as are found generally in the north of Europe, and referred by geologists to what they call "the Glacial Epoch." There are not, however, wanting abundant evidences of the transport of drift materials, and there is some good evidence, collected by Captain M'Clintock, of the direction in which the drift was moved.

Specimens of granite, which I have no hesitation in referring to the characteristic granite of the west side of North Somerset, were found at Leopold Harbour (North Somerset) and at Graham Moore Bay (Bathurst Island); one of these localities is N.E. and the other N.W. of the granite of North Somerset, from which I infer that there was no constant prevailing direction for the drift ice that carried these boulders, but that they were transported to the northward in various directions, according to the varying motion of the currents that moved the ice. The boulder of granite at Port Leopold is 100 miles N.E. of the granite which gave origin to it; and the specimens from Graham Moore Bay are 190 miles to the N.W. of their source.

At Cape Rennell (North Somerset), in a direction intermediate between the two former directions, a remarkable boulder of the same granite was found, confirming the general direction of the transporting force from south to north. Its position and size are thus recorded by Captain M'Clintock:—" Near Cape Rennell we passed a very remarkable rounded boulder of gneiss or granite; it was 6 yards in circumference, and stood near the beach, and some 15 or 20 yards above it; one or two masses of rounded gneiss, although very much smaller, had arrested our attention at Port Leopold."

It is well known that Captain Sir Robert M'Clure brought home specimens of pine-trees found in the greatest abundance in the ravines on the west coast of Baring Island; one of his specimens preserved in the museum of the Royal Dublin Society measures 15 inches by 12 inches, and contains three knots that prove it formed a portion of the stem high above its root. The bark is not found on this specimen, which does not represent the full thickness of the tree; I have estimated that this fragment contains 70 rings of annual growth.

Similar remains were found by Captain M'Clintock and Lieutenant Mecham in Prince Patrick's Island, and in Wellington Channel by Sir Edward Belcher. On the coast of New Siberia, Lieutenant Anjou found a clay cliff containing stems of trees still capable of being used as fuel. The original observers all agree in thinking that these trees grew where they are now found; and Captain Osborn, in mentioning Sir Roderick I. Murchison's opinion that they are drift timber, justly adds the remark, that a sea sufficiently free from ice to allow of their being drifted from the south would indicate also a climate sufficiently mild to allow of their having grown upon the land where they now occur. Mr. Hopkins, in his anniversary address as President of the Geological Society of London, has published a remarkable geological speculation, which would account for the facts above mentioned.* So far as the evidence of drift boulders is concerned, I have shown that the

* Journ. Geol. Soc. Lond., vol. VIII. p. lxiv.

direction of the currents was from the south; a fact which falls in with the drift theory, so far as it goes.

We cannot, however, dissociate these trees from the facts connected with the distribution of the remains of the Siberian Mammoth in Asia and America. It is now known that this elephant was provided with a warm fur, and that his food was of a kind which grows even now in Northern Siberia; so that the drift theory, which was formerly supposed necessary to account for the occurrence of these remains, has now been quietly dropped, *sub silentio,* by the geologists. Many other drift theories have, in like manner, lived their short day, and gone the way of all false hypotheses; among others, the drift theory of the origin of coal. Further investigation may show that the glacial epoch of Europe was one of a very different character in Asia and America, and that, while glaciers clothed the sides of Snowdon and Lugnaquillia, pine forests flourished in the Parry Islands, and the Siberian elephants wandered on the shores of a sea washed by the waves of an ocean that carried no drifting ice.

There is abundant evidence, however, that the Arctic Archipelago was submerged in very recent geological periods; for we know that subfossil shells, of species that now inhabit the waters of the neighbouring seas, are found at considerable heights throughout the whole group of islands. M'Clure found shells of the *Cyprina Islandica* at the summit of the Coxcomb range, in Baring Island, at an elevation of 500 feet above the sea-level; Captain Parry, also, has recorded the occurrence of *Venus* (probably *Cyprina Islandica*) on Byam Martin's Island; and in the recent voyage of the 'Fox,' Dr. Walker, the surgeon of the expedition, found the following subfossil shells at Port Kenedy, at elevations of from 100 to 500 feet:—

1. *Saxicava rugosa.*
2. *Tellina proxima.*
3. *Astarte Arctica* (Borealis).
4. *Mya Uddevallensis.*
5. *Mya truncata.*
6. *Cardium* sp.
7. *Buccinum undatum.*
8. *Acmea testudinalis.*
9. *Balanus Uddevallensis.*

At the same place a portion of the palate-bone of a whale (Right Whale) was found at an elevation of 150 feet.

All these facts indicate the former submergence of the Arctic Archipelago, but this submergence must have been anterior to the period when pine forests clothed the low sandy shores of the slowly emerging islands, the remains of which forests now occupy a position at least 100 feet above high-water mark.

The geological map which I am enabled to publish from the data collected by Captains M'Clintock, M'Clure, Osborn, &c., is an enlargement of that which was published in 1857 by the Royal Society of Dublin, to illustrate the fine collection of Arctic fossils and minerals deposited in the museum of that body by Captains M'Clintock and M'Clure. In perfecting it for its present purpose I have availed myself of all the other sources of information within my reach, among which I am bound to mention in particular the excellent Appendix to Dr. Sutherland's 'Voyage of the Lady Franklin and Sophia,' written by Mr. Salter, Palæontologist of the Geological Survey of Great Britain.

Many of the mineral specimens from Greenland, and the fossils from Cape Riley, Cape Farrand, Point Fury, and Brentford Bay, were collected by Dr. David Walker, surgeon and naturalist to the 'Fox' Expedition.

No. V.

LIST OF SUBSCRIBERS TO THE 'FOX' EXPEDITION.

	£.	s.	d.
ACLAND, Sir T. D., Bart. ..	100	0	0
Adams, Dr. Walter, Edinburgh	3	3	0
Aldrich, Captain, R.N.	1	1	0
Allan, Rob. M., Esq.	1	1	0
Allen, Captain Robert	5	5	0
Allen, Captain, R.N.	2	2	0
Ames, Mrs.	5	0	0
Ames, Miss	1	0	0
Anon.	5	0	0
Armstrong, Mrs.	1	1	0
Armstrong, children of Mrs...	0	8	9
Arnold, Mrs.	1	1	0
Arrowsmith, John, Esq. ..	5	0	0
Austin, Rear-Adm. Horatio T., R.N., C.B.	5	0	0
BABBAGE, Charles, Esq. ..	10	0	0
Baikie, Dr.	1	1	0
Baker, Mrs.	5	0	0
Barkworth, Geo., Esq.	5	0	0
Barras, Miss	1	1	0
Barrett, H. J., Esq.	1	0	0
Barrow, John, Esq.	25	0	0
Barstow, Lieutenant, R.N. ..	1	0	0
Barth, Dr. Henry	5	5	0
Bath, W. J. C., Esq.	0	2	6
Batty, Mrs. J. M.	1	1	0
Beaufort, Rear-Adm. Sir Francis, K.C.B.	50	0	0
Bell, Thos., Esq., Pres. Lin. Soc.	10	10	0
Bennett, John S., Esq.	5	0	0
Birch, J. W. N., Esq.	10	0	0
Bird, Captain, R.N.	5	0	0
Birmingham, small sums collected at Evans's Library ..	3	1	0
Booth, Mrs.	5	0	0
Borton, Mrs., collected by ..	1	10	0
Boston, coll. at, by Mr. Morton	4	4	0
Bovill, Walter, Esq.	5	0	0
Boyer, Lieut. R.N.	0	10	0
Boyle, the Hon. Carolina C.	1	0	0
Brigg, collected at	1	1	0
Brine, Captain, R.E.	1	1	0
Brooking, J. Holdsworth, Esq.	10	0	0
Brown, Robert, Esq., V.P.L.S.	20	0	0
Brown, John, Esq.	5	5	0
Brown, J. E., Esq., R.N. ..	0	5	0
Bruce, the Rev. C.	1	1	0
Burgoyne, Captain, R.N. ..	1	0	0
Burton, Alfred, Esq.	1	1	0
Byron, the Hon. Fred.	5	0	0
CHESNEY, Major-General ..	2	2	0
Collinson, Captain, R.N., C.B.	20	0	0
Coningham, W. Esq., M.P. ..	100	0	0
Coote, C. W., Esq.	1	0	0
Coote, Charles, Esq.	10	0	0
Courtauld, Samuel, Esq. ..	25	0	0
Courtauld, George, Esq. ..	15	0	0
Coutts, Messrs., & Co.	50	0	0
Crasp, J., Esq., Surgeon, 63rd Regt.	1	0	0
Crauford, John, Esq.	5	0	0
Cresswell, S. Gurney, Commander, R.N.	5	0	0
DALGETY, F. T., Esq.	10	10	0
De la Roquette, M., V. P. of Geog. Soc. of Paris, 1000 fr.	40	0	0
Dilke, C. W., Esq.	5	0	0
Dixon, James, Esq.	10	0	0
Doxat, Alexis J., Esq.	10	10	0
Doxat, Miss H., collected by ..	4	0	0
"Dubious"	0	2	6
Dufferin, Lord	25	0	0
EDGAR, Mrs., collected by ..	5	0	0
Ellesmere, the Earl of	15	0	0
Elphinstone, the Hon. Mount-Stewart	10	0	0
Elton, Sir Arthur H., Bart. ..	5	5	0
Emanuel, Ezekiel, Esq.	1	0	0
FAIRHOLME, the Hon. Mrs. ..	150	0	0
Filliter, George, Esq.	10	0	0
Fitton, Dr.	21	0	0
Fortescue, Rev. T. F. G. ..	2	2	0
GARLING, H., Esq.	1	1	0
Gassiot, J. P., Esq.	25	0	0
Gimingham, W., Esq., & Mrs.	2	2	0
Gipps, Lady	5	0	0
Gowen, J. R., Esq.	5	0	0
Graves, Messrs., Pall Mall ..	1	1	0
Griffiths, G. H., Esq.	5	5	0
Gruneisen, Ch. Lewis, Esq. ..	1	1	0
Gruneisen, Mrs.	1	1	0
Guillemard, the Rev. W. H. ..	5	0	0
Guillemard, Miss	1	0	0
HALL, Jas., Esq.	5	0	0
Hanbury, Mrs.	1	1	0
Hardinge, Commander, R. N.	0	10	0
Hardwicke, Philip, Esq. ..	5	0	0
Harney, Julian, Esq., collected by, at Jersey	50	0	0
Heales, Alfred, Esq.	5	5	0
Herring, Miss	2	2	0

	£.	s.	d.
Hicks, John, Esq...	2	0	0
Hill, Col., 63rd Regt.	1	0	0
Hodgson, Mrs.	10	0	0
Holland, Commander, R. N...	5	0	0
Hollingsworth, H., Esq. ..	2	2	0
Hollond, Rob., Esq.	10	10	0
Hooker, Dr. J. D...	5	5	0
Hornby, Miss Georgina ..	100	0	0
Hornby, the Rev. Edward ..	25	0	0
Hornby, Mrs. Edmund.. ..	5	0	0
Hornby, Miss Georgina, collected by	13	4	0
Hovell, W. H., Esq.	5	5	0
Hughes, Lieutenant, R.N. ..	2	0	0
Inglis, Lady	10	0	0
Irby, T. W., Esq.	1	1	0
Jackson, N. Ward, Esq. ..	21	0	0
Janson, J. C., Esq.	5	5	0
Jeans, H. W., Esq., R.N. ..	0	10	0
Jersey "Times"	2	10	0
Kellett, Commodore, C.B.	10	0	0
Kendall, Mrs.	1	0	0
Kendall, the Rev. Professor ..	1	0	0
Key, Lieut., R.N..	0	5	0
King, William, Esq.	5	0	0
Laird, Macgregor, Esq. ..	50	0	0
Laird, John, Esq..	25	0	0
L. and N. W.	1	4	0
Lanford, J., Esq., Quartermaster 63rd Regiment ..	0	10	0
Langhorne, A., Esq.	1	1	0
Larcom, Mrs.	1	0	0
Leach, William, Esq.	5	5	0
Le Feuvre, W. J., Esq. ..	50	0	0
Lefroy, C. E., Esq.	2	0	0
Leicester, the Rev. F.	1	1	0
Lethbridge, Lieut., R.N. ..	0	5	0
"Lochmaben Castle," Owners of the	5	5	0
Lyall, D. Esq., R.N., M.D. ..	5	0	0
Mackintosh, Eneas, Esq. ..	10	0	0
Maguire, Captain, R.N. ..	3	3	0
Maitland, Capt. Sir Thos., R.N.	1	0	0
Majendie, Ashhurst, Esq., and Mrs.	100	0	0
Servants of the above ..	0	14	0
Malby, Messrs.	5	0	0
Malby, Messrs., Workmen in their Establishment by a 6*d*. Subscription	4	11	6
Mansfield, W. H. S., Esq. ..	0	10	0
Mantell, Dr. A. A.	1	0	0
Markham, Clements, Esq. ..	1	1	0
Markham, Mrs.	1	0	0
M'Crea, Captain, R.N.	0	10	0

	£.	s.	d.
M'Kinlay, Miss	1	0	0
M'Kinlay, Miss Elizabeth ..	1	0	0
M'William, Dr., R.N.	1	1	0
Merry, W. L., Esq.	1	1	0
Morris, Rev. F. B.	1	0	0
Morris, Sir Armine, Bart. ..	5	0	0
Murchison, Sir Roderick Impey, G.C.St.S., President of the Royal Geographical Society	100	0	0
Murray, John, Esq.	20	0	0
Nares, Fras., Esq.	2	2	0
Newall, W. L., Esq.	100	0	0
Nicholson, Sir Charles.. ..	5	0	0
N. J.	2	2	0
Norwood, collected at, by a Lady	7	15	0
Ommanney, Capt. Erasmus, R.N.	2	0	0
Osborn, Sir George, Bart. ..	1	0	0
Paget, A. F., Esq.	0	10	6
Paget, C. H. M., Esq.	1	1	0
Pasley, Gen. Sir Charles W., K.C.B.	10	0	0
Second Subscription	10	0	0
Third Subscription	5	0	0
Pattinson, H. L., Esq... ..	50	0	0
Pearce, Stephen, Esq.	2	2	0
Phillimore, Captain, R.N. ..	2	2	0
Pigou, Fred., Esq...	10	0	0
Prescott, Vice-Admiral Sir Henry, K.C.B...	5	0	0
Rawnsley, the Rev. Drummond..	5	0	0
Rawnsley, Mrs., collected by	1	0	0
Rawnsley, Willingham Franklin, collected by, at Uppingham School	0	10	0
Raynsford, Mrs..	1	1	0
Reynardson, H. B., Esq. ..	5	0	0
Rogers, Lieut., R.N.	1	0	0
Roget, Dr. P. M.	5	0	0
Roper, Geo., Esq...	5	5	0
Ross, Rear-Admiral Sir Jas. C.	21	0	0
Rupert's Land, Bishop of ..	5	0	0
Sabine, Major-General ..	25	0	0
Sadler, W. F., Esq.	10	10	0
Sefton, the Countess of ..	10	0	0
Shearley, W., Esq.	2	0	0
Sheil, Sir Justin	5	0	0
Shewell, John Tulmin, Esq.	5	5	0
Simpson, J., Esq., R.N. ..	1	10	0
Skey, Dr.	2	2	0
Smith, Eric E., Esq.	2	0	0
Smith, John Henry, Esq. ..	10	10	0
Smith, Osborne, Esq... ..	2	2	0

	£.	s.	
Smith, Archibald, Esq. ..	5	5	
Sparrow, Jas., Esq.	5	0	0
St. Asaph, the Bishop of ..	10	0	0
St. David's, the Bishop of ..	10	0	0
St. Leger, A. B.	5	0	0
Stainton, J. J., Esq.	3	3	0
Statham, J. L., Esq.	1	1	0
Stephenson, Robert, Esq. ..	20	0	0
Stirling, Commander, R.N. ..	0	10	0
Strzelecki, Count P. de ..	25	0	0
Swinburne, Rear-Admiral ..	30	0	0
Sykes, Col., M.P.	5	0	0
TAYLOR, William, Esq. ..	5	0	0
Tennant, James, Esq.	2	0	0
T. H., collected in shillings by	2	0	0
Thackeray, W. M., Esq. ..	5	0	0
Thomson, J., Esq.	1	1	0
Tindal, Commander, R.N. ..	2	2	0
Tinney, W. H., Esq., Q.C. ..	20	0	0
Tite, W., Esq., M.P.	50	0	0
Trevelyan, Sir W. C., Bart...	40	0	0
Trevelyan, Lady	10	0	0
Trevilian, M. C., Esq.	2	2	0
Trollope, Commander, R.N...	2	2	0
Tuckett, Fred., Esq.	5	0	0
Tudor, J., Esq.	0	10	0
Turner, Alfred, Esq.	15	0	0
Tweedie, W. M., Esq.	5	0	0
VINCENT, John, Esq.	1	0	0
WALKER, James, Esq. ..	21	0	0

	£.	s.	d.
Washington, Captain, R.N., Hydrographer of the Navy	21	0	0
Waterfield, Edward, Esq. ..	5	0	0
Wayse, the Rev. J. W. ..	5	0	0
Weld, Charles R., Esq. ..	5	0	0
Wheatstone, Professor	5	0	0
Willes, Hon. Mr. Justice ..	21	0	0
Wilson, Robert, Esq.	1	1	0
Wittenoom, Miss	1	1	0
Wodehouse, Commander ..	0	10	0
Woodcock, J. Parry, Esq. ..	5	0	0
Worsley, Marcus, Esq.	10	0	0
Wright, the Rev. R. F. ..	2	2	0
Wrottesley, Lord	50	0	0
YOUNG, Chas. F., Esq. ..	5	0	0
Young, Miss	5	0	0
Young, A. Verity, Esq. ..	2	2	0
Yule, Mrs. H.	5	0	0
The brother and sisters of the late John and Thomas Hartnell, of H.M.S. 'Erebus,' buried at Beechey Island ..	5	0	0
A Commander R.N.	0	5	0
A Commander in the Merchant Service	500	0	0
A Friend. C. H.	5	0	0
A Friend	1	0	0
The daughters of a retired Commander	2	0	0
A Sympathiser	1	0	0
	£2981	8	9

A life-boat, presented by Messrs. White of Cowes.

A large quantity of preserved potatoes, by Messrs. King, late Edwards.

Apparatus for lowering a boat at sea, presented by Mr. Clifford, the inventor.

Three travelling-tents, by Messrs. Winsor and Newton.

A stove, by Mr. Rettie.

20 dozen "Isle of Wight Sauce," by Mr. Tucker of Newport.

Apparatus for reefing topsails, from Mr. Cunningham, the inventor.

EXPENSES OF THE EXPEDITION.

	£.	s.	d.
Purchase of the 'Fox' steam yacht	2,000	0	0
Strengthening and refitting for Arctic service	1,666	15	7
Engine repairs and alterations	450	0	0
Engine stores	256	19	9
Provisions	1,374	16	7
Clothing	240	10	6
Sundries for the use of the Expedition	189	15	5
Aberdeen Steam Company, for carriage of stores and passage of crew	68	13	0
Provisions, dogs, fuel, &c., in Greenland	123	0	6
Provisions purchased from the Whaler 'Emma,' in Baffin's Bay	36	2	5
Pay and wages to officers and crew, including allotments to their wives and families during the absence of the Expedition *	3,888	2	9
Pilotage, boat-hire, ship-keeper, dock-labour, &c.	34	11	0
Carriage of boat from Liverpool	33	15	0
Miscellaneous, including printers' bills, advertisements, telegrams, legal expenses, &c.	49	16	6
	£10,412	19	0

The above expenses of the Expedition would have been considerably increased, but for the great liberality of Messrs. Bayley and Ridley, of Cooper's Court; of the Directors of the East and West India Dock Company; of Messrs. Richard and Henry Green, Blackwall; of Messrs. T. and W. Smith, of the Royal Exchange Buildings; of Messrs. Forest, of Limehouse; and of Mr. Westhrop, of Poplar; all of whom placed their establishments at the service of Lady Franklin on the return of the 'Fox,' and declined receiving any remuneration whatever.

* The crew of the 'Fox' received the usual double pay, granted by the Admiralty to all employed in Arctic service.

LONDON: PRINTED BY W. CLOWES AND SONS, STAMFORD STREET, AND CHARING CROSS.

ALBEMARLE STREET,
January, 1860.

MR. MURRAY'S

LIST OF NEW WORKS.

THE CIVIL CORRESPONDENCE AND MEMORANDA OF FIELD-MARSHAL THE DUKE OF WELLINGTON WHILE CHIEF SECRETARY FOR IRELAND, from 1807 to 1809. 8vo. 20*s*.

ON THE ORIGIN OF SPECIES BY MEANS OF NATURAL SELECTION. By CHARLES DARWIN, M.A., Author of 'Naturalist's Voyage Round the World.' Post 8vo. 14*s*.

LIFE AND JOURNALS OF THE RIGHT REV. DANIEL WILSON, D.D., late Lord Bishop of Calcutta. By Rev. JOSIAH BATEMAN, M.A. Portraits and Illustrations. 2 vols. 8vo. 28*s*.

THOUGHTS ON GOVERNMENT AND LEGISLATION. By LORD WROTTESLEY, F.R.S. Post 8vo. 7*s*. 6*d*.

HISTORICAL EVIDENCES OF THE TRUTH OF THE SCRIPTURE RECORDS STATED ANEW, with Special Reference to the Doubts and Discoveries of Modern Times. By Rev. GEORGE RAWLINSON, M.A. 8vo. 14*s*.

THE ARCHÆOLOGY OF BERKSHIRE; BY LORD CARNARVON. Second Edition. Post 8vo. 1*s*.

THE STORY OF NEW ZEALAND; PAST AND PRESENT—SAVAGE AND CIVILISED. By ARTHUR S. THOMSON, M.D., 58th Regiment. Illustrations. 2 vols. Post 8vo. 24*s*.

MODERN SYSTEMS OF FORTIFICATION, Examined with reference to the NAVAL, LITTORAL, and INTERNAL DEFENCE of ENGLAND. By GENERAL SIR HOWARD DOUGLAS, Bart. Plans. 8vo. 12*s*.

SCIENCE IN THEOLOGY. Sermons preached before the University of Oxford. By Rev. ADAM S. FARRAR, Fellow of Queen's College. 8vo. 9*s.*

BECKET; A BIOGRAPHY. By Rev. CANON ROBERTSON, M.A. Illustrations. Post 8vo. 9*s.*

ON THE INTUITIVE CONVICTIONS OF THE MIND. By Rev. JAMES M'COSH, LL.D., Professor in Queen's College, Belfast. 8vo.

MEMOIRS OF THE EARLY LIFE OF LORD CHANCELLOR SHAFTESBURY. With his Letters, Speeches, &c. By W. D. CHRISTIE. Portrait. 8vo. 10*s.* 6*d.*

SELF HELP. With Illustrations of Character and Conduct. By SAMUEL SMILES, Author of 'Life of George Stephenson.' Post 8vo. 6*s.*

A DICTIONARY OF BIBLICAL ANTIQUITIES. BIOGRAPHY, GEOGRAPHY, AND NATURAL HISTORY. Edited by WILLIAM SMITH, LL.D. Woodcuts. Vol. I. Med. 8vo. 42*s.* (*Just ready.*)

LIFE AND TIMES OF THE PIOUS ROBERT NELSON, Author of 'Companion to the Fasts and Festivals of the Church.' By Rev. C. F. SECRETAN, M.A. Portrait. 8vo.

MANNERS AND CUSTOMS OF THE MODERN EGYPTIANS. By E. W. LANE. A New Edition, with Additions. Edited by E. STANLEY POOLE. Woodcuts. 8vo. 18*s.*

PICTURES OF THE CHINESE. Drawn by Themselves. Described by Rev. R. H. COBBOLD, M.A., late Archdeacon of Ningpo. 34 Plates. Crown 8vo. 9*s.*

A MANUAL OF THE BRITISH CONSTITU- TION; being a Review of its Rise, Growth, and Present State. By DAVID ROWLAND. Post 8vo. 10*s.* 6*d.*

MEMOIRS ofthe **EARLY ITALIAN PAINTERS,** AND OF THE PROGRESS OF PAINTING IN ITALY. By MRS. JAMESON. New Edition, with Additions. Woodcuts. Post 8vo. 12*s.*

EOTHEN; Or, Traces of Travel brought Home from the East. New Edition, with Frontispiece. Post 8vo. 7*s.* 6*d.*

A MEMOIR OF PATRICK FRASER TYTLER, the Historian of Scotland. By Rev. J. W. BURGON, M.A. Second Edition. Crown 8vo. 9*s*.

THE ITALIAN VALLEYS OF THE ALPS; a Tour through all the Romantic and less frequented "Vals" of Northern Piedmont. By Rev. S. W. KING. Illustrations. Crown 8vo. 18*s*.

THE MARQUIS CORNWALLIS'S PAPERS AND CORRESPONDENCE during his Administrations in India, America, Ireland, &c. Edited by CHARLES ROSS. Second Edition, revised. Portrait. 3 vols. 8vo. 63*s*.

THE STORY OF THE LIFE OF GEORGE STEPHENSON. By SAMUEL SMILES. Illustrations. 7th Thousand. Post 8vo. 6*s*.

THREE VISITS TO MADAGASCAR. With Notices of the People, Natural History, &c. By Rev. W. ELLIS. 5th Thousand. Illustrations. 8vo. 16*s*.

SERMONS PREACHED FOR THE MOST PART IN CANTERBURY CATHEDRAL. By Rev. CANON STANLEY, D.D. Post 8vo. 7*s*. 6*d*.

ON COLOUR; and on the Necessity for a General Diffusion of Taste among all Classes. By Sir J. G. WILKINSON. Illustrations. 8vo. 18*s*.

LIFE OF JAMES WATT. With Selections from his Correspondence. By JAMES P. MUIRHEAD, M.A. Second Edition. Portrait, &c. 8vo.

REMARKS ON ITALY, during Several Visits from 1816 to 1854. By LORD BROUGHTON. Second Edition. 2 vols. Post 8vo. 18*s*.

SILURIA: The History of the Oldest Fossiliferous Rocks and their Foundations. By SIR R. MURCHISON. Third Edition. Illustrations. 8vo. 42*s*.

HISTORICAL ACCOUNT OF THE FOREST OF DEAN. By Rev. H. G. NICHOLLS. Woodcuts, &c. Post 8vo. 10*s*. 6*d*.

JOHN MURRAY, ALBEMARLE STREET.

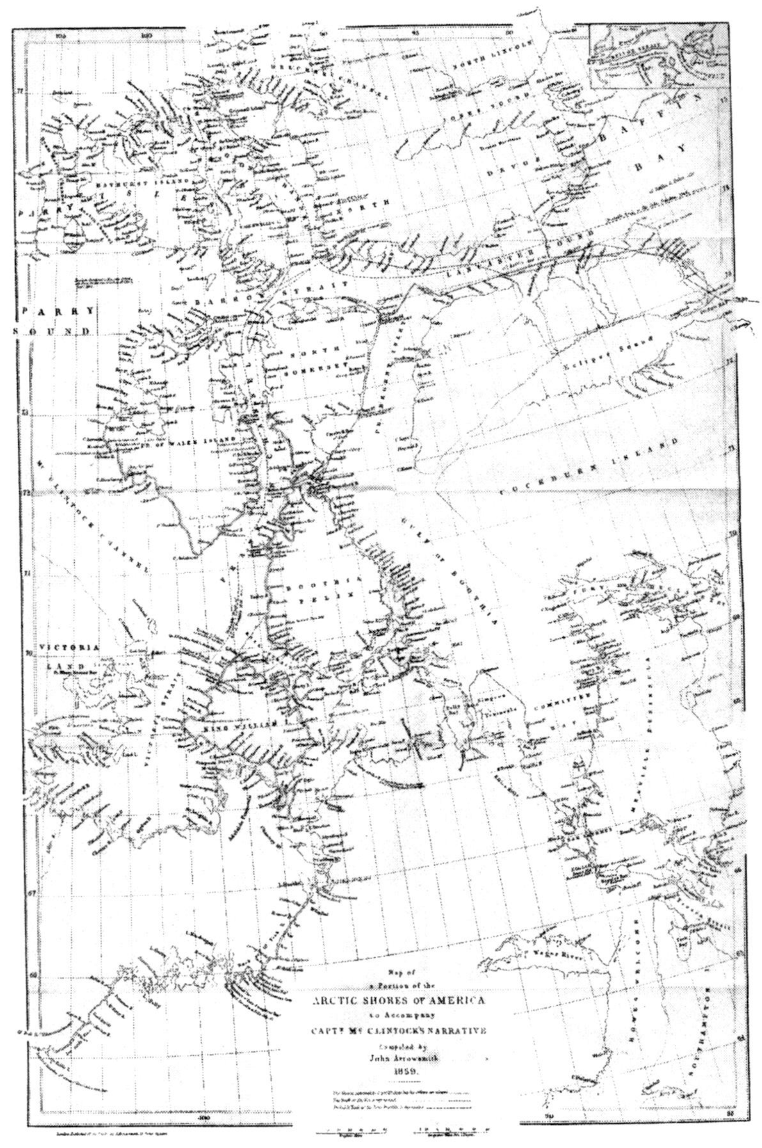

Map of
a Portion of the
ARCTIC SHORES OF AMERICA
to Accompany
CAPTN. MC CLINTOCK'S NARRATIVE
Compiled by
John Arrowsmith
1859.
BAFFIN BAY
LANCASTER SOUND
BARROW STRAIT
NORTH SOMERSET
PARRY SOUND
PARRY ISLES
BATHURST ISLAND
NORTH LINCOLN
JONES SOUND
DEVON
NORTH
PR. OF WALES ISLAND
COCKBURN ISLAND
Eclipse Sound
MC CLINTOCK CHANNEL
BOOTHIA FELIX
GULF OF BOOTHIA
VICTORIA LAND
KING WILLIAM
COMMITTEE BAY
MELVILLE PENINSULA
ROWES WELCOME
SOUTHAMPTON
Wager River

Made in the USA